EXPLAINING PRIMARY SCIENCE

Sara Miller McCune founded SAGE Publishing in 1965 to support the dissemination of usable knowledge and educate a global community. SAGE publishes more than 1000 journals and over 800 new books each year, spanning a wide range of subject areas. Our growing selection of library products includes archives, data, case studies and video. SAGE remains majority owned by our founder and after her lifetime will become owned by a charitable trust that secures the company's continued independence.

Los Angeles | London | New Delhi | Singapore | Washington DC | Melbourne

EXPLAINING PRIMARY SCIENCE

PAUL CHAMBERS
NICHOLAS SOUTER

INCLUDES
CLASSROOM
EXPERIMENT
VIDEOS

⑤SAGE

Los Angeles | London | New Delhi
Singapore | Washington DC | Melbourne

Los Angeles | London | New Delhi
Singapore | Washington DC | Melbourne

SAGE Publications Ltd
1 Oliver's Yard
55 City Road
London EC1Y 1SP

SAGE Publications Inc.
2455 Teller Road
Thousand Oaks, California 91320

SAGE Publications India Pvt Ltd
B 1/I 1 Mohan Cooperative Industrial Area
Mathura Road
New Delhi 110 044

SAGE Publications Asia-Pacific Pte Ltd
3 Church Street
#10-04 Samsung Hub
Singapore 049483

Editor: James Clark
Assistant editor: Robert Patterson
Production editor: Nicola Marshall
Copyeditor: Audrey Scriven
Proofreader: Thea Watson
Indexer: David Rudeforth
Marketing manager: Lorna Patkai
Cover design: Sheila Tong
Typeset by: C&M Digitals (P) Ltd, Chennai, India
Printed and bound in the UK by Ashford Colour
Press Ltd.

Library of Congress Control Number: 2016949499

British Library Cataloguing in Publication data

A catalogue record for this book is available from
the British Library

ISBN 978-1-4739-1279-3
ISBN 978-1-4739-1280-9 (pbk)

TABLE OF CONTENTS

About the Authors   vii
Curriculum Links – England   ix
Curriculum Links – Scotland   xvii
Companion Website   xxiii

1   Biodiversity   1

2   Ecosystems   27

3   Cell Biology   57

4   Plants   87

5   Life Processes   121

6   Inheritance, Genes and Life   145

7   Animal Behaviour   169

8   Types of Matter   189

9   Properties and Uses of Matter   207

10 Materials from the Earth   221

11 Chemical Changes   235

12 Water and its Uses   247

13 Space   263

14 Energy   281

15   Electricity   293

16   Light   309

17   Sound   325

18   Forces   341

19   Gravity and Weight   357

Bibliography   373
Index   375

ABOUT THE AUTHORS

Paul Chambers worked in industry prior to attending Jordanhill College of Education where he qualified as a physics and mathematics teacher. He then taught in a number of schools and was head of physics for over fourteen years. Whilst at school he was seconded on a number of occasions as staff trainer in physics and science. He was appointed lecturer in physics education (Initial Teacher Education) at the University of Strathclyde in 1999 and senior lecturer in 2003. He was also head of department for STEM education from 2003–2005 and a member of the Physics Qualifications Development Team which produced the new physics curricula in Scotland. He is the author of over ten school texts, from upper primary through to upper secondary, and has contributed to national examinations in physics as a marker, setter and vetter.

More recently he has been developing science materials and CPD for primary schools and is co-ordinator for all STEM activities at the School of Education at the University of Strathclyde. He has taught on teacher education courses in Egypt and Malawi and was External Examiner for PGCE Science at an English university. He is also an accredited provider of primary science CPD for the Scottish Schools Education Research Centre (SSERC), and serves on the ASE's *Primary Science* editorial board.

Nicholas (Nicky) Souter's first science lesson in 'the big school' included being told that the universe consists of matter and energy. That elegant and intriguing starting point was the hook that set him on a lifelong journey of sharing science with school pupils and teachers. He possesses undergraduate and postgraduate qualifications in life science and in education awarded by the University of Glasgow, Jordanhill College of Education, Paisley College of Technology and the Royal Society of Biology. He established a reputation as an innovative secondary science practitioner and leader while developing a profile in curriculum development, teacher's CPD and as an examiner in biology. He moved midcareer into full-time teacher education at the University of Strathclyde where he progressed to senior lecturer as well as holding a number of senior faculty positions including leadership for one of the largest PGCE programmes in Europe, external examining, and the accreditation and design of new teacher education programmes in England and Scotland. Nicky has been a member of a number of committees in professional organisations e.g. The Association for Science Education, the Society of Biology and the British Ecological Society. He has undertaken consultancies in Pakistan and Germany, has contributed to more than 25 books, several of which offered supportive advice to life science teaching at all levels of the school curriculum. He is also an accredited provider of primary science CPD for SSERC, and has held consultancies with Rolls-Royce, the Woodland Trust, OXFAM and the BBC. Nicky is a Chartered Science Teacher and a Fellow of the Royal Society of Biology.

CURRiCULUM LiNKS - ENGLAND

The following table shows how chapter coverage links to the statutory requirements of the National Curriculum in England (DfE, 2013).

Working scientifically	Relevant chapters
Pupils should be taught to use the following practical scientific methods, processes and skills via the teaching of the programme of study content:	
• asking simple questions and recognising that they can be answered in different ways (Y1&2).	1, 2, 5, **6**, 8, 9, 10, 11, 12, 13, 14, 15, 16, 17, 18, 19
• observing closely, using simple equipment (Y1&2).	**1**, 2, **3**, 4, 5, **6**, 11, 12, 15, 16, 17, 18, 19
• performing simple tests (Y1&2).	5, 6, **7**, **12**, **16**, **17**, **18**, **19**
• identifying and classifying (Y1&2).	**1**, **2**, **4**, 5, **6**, 7, 10, 13
• using their observations and ideas to suggest answers to questions (Y1&2).	2, 3, 4, 5, **6**, 11, 12, 15, 16, 17, 18, 19
• gathering and recording data to help in answering questions (Y1&2).	1, 2, **3**, 4, **5**, **6**, 7, 8, 9, 10, 11, 12, 13, 14, 15, 16, 17
• asking relevant questions and using different types of scientific enquiries to answer these (Y3&4).	1, **6**, **10**, **12**, **13**, **14**, **15**, **16**, **17**, **18**, **19**
• setting up simple practical enquiries, comparative and fair tests (Y3&4).	1, 2, **3**, 4, **5**, **6**, **7**, 8, 9, 10, 11, 12, 13, 14, 15, 16, 17, 18, 19
• making systematic and careful observations, and where appropriate taking accurate measurements using standard units and a range of equipment, including thermometers and data loggers (Y3&4).	**1**, 2, **3**, 4, 5, **6**, 7, 11, 14, 15
• gathering, recording, classifying and presenting data in a variety of ways to help in answering questions (Y3&4).	1, 2, 3, **5**, 6, 7, 11, 18, 19
• recording findings using simple scientific language, drawings, labelled diagrams, keys, bar charts, and tables (Y3&4).	1, **2**, 3, 4, **5**, **6**, 7, 8, 9, 10, 11, 12, 13, 14, 15, 16, 17, 18, 19

[bold] – indicate particular/extended coverage
[non-bold] – indicate the targets are threaded throughout the chapter.

(Continued)

• reporting on findings from enquiries, including oral and written explanations, displays or presentations of results and conclusions (Y3&4).	1, **2**, 3, 4, **5**, **6**, **7**, 8, 9, 10, 11, 12, 13, 14, 15, 16, 17, 18, 19
• using results to draw simple conclusions, make predictions for new values, suggest improvements, and raise further questions (Y3&4).	1, 2, 3, 4, 5, 6, 7, 8, 9, 10, 11, 12, 13, 14, 15, 16, 17, 18, 19
• identifying differences, similarities or changes related to simple scientific ideas and processes (Y3&4).	**1**, 5, **6**, 11
• using straightforward scientific evidence to answer questions or support their findings (Y3&4).	1, 2, **6**, 7, **11**, **15**, **16**, **17**, **18**, **19**
• planning different types of scientific enquiries to answer questions, including recognising and controlling variables where necessary (Y5&6).	**1**, **2**, **3**, 4, **5**, **6**, 7, 8, 9, 10, 11, 12, 13, 14, 15, 16, 17, 18, 19
• taking measurements, using a range of scientific equipment, with increasing accuracy and precision, and taking repeat readings when appropriate (Y5&6).	**1**, **2**, **5**, **6**, **11**, **12**, **15**, 16, 17, **18**, **19**
• recording data and results of increasing complexity using scientific diagrams and labels, classification keys, tables, scatter graphs, bar and line graphs (Y5&6).	**1**, **5**, **6**, 8, 10, 11, 12, 15, 16, 17, 18, 19
• using test results to make predictions to set up further comparative and fair tests (Y5&6).	2, 3, **5**, **6**, 8, 10, 11, 15, 16, 17, 18, 19
• reporting and presenting findings from enquiries, including conclusions, causal relationships and explanations of and degree of trust in results, in oral and written forms such as displays and other presentations (Y5&6).	1, 3, **5**, **6**, 7, 11, 15, 16, 17, 18, 19
• identifying scientific evidence that has been used to support or refute ideas or arguments (Y5&6).	**1**, 2, **6**, 7, **14**, **18**, **19**

Plants	**Relevant chapters**
Pupils should be taught to:	
• identify and name a variety of common wild and garden plants, including deciduous and evergreen trees (Y1).	**1**, 2, **4**, **6**, 7
• identify and describe the basic structure of a variety of common flowering plants, including trees (Y1).	4
• observe and describe how seeds and bulbs grow into mature plants (Y2).	**1**, **4**
• find out and describe how plants need water, light and a suitable temperature to grow and stay healthy (Y2).	4
• identify and describe the functions of different parts of flowering plants: roots, stem/trunk, leaves and flowers (Y3).	4
• explore the requirements of plants for life and growth (air, light, water, nutrients from soil, and room to grow) and how they vary from plant to plant (Y3).	**1**, 2, 3, **4**, 6
• investigate the way in which water is transported within plants (Y3).	4
• explore the part that flowers play in the life cycle of flowering plants, including pollination, seed formation and seed dispersal (Y3).	**1**, 3, **4**

[bold] – indicate particular/extended coverage
[non-bold] – indicate the targets are threaded throughout the chapter.

Animals including humans	Relevant chapters
Pupils should be taught to:	
• identify and name a variety of common animals including fish, amphibians, reptiles, birds and mammals (Y1).	1
• identify and name a variety of common animals that are carnivores, herbivores and omnivores (Y1).	**2**, 5
• describe and compare the structure of a variety of common animals (fish, amphibians, reptiles, birds and mammals, including pets) (Y1).	5
• identify, name, draw and label the basic parts of the human body and say which part of the body is associated with each sense (Y1).	5, **7**
• notice that animals, including humans, have offspring which grow into adults (Y2).	**3**, **5**, **6**, 7
• find out about and describe the basic needs of animals, including humans, for survival (water, food and air) (Y2).	**2**, 5, **7**
• describe the importance for humans of exercise, eating the right amounts of different types of food and hygiene (Y2).	2, **3**, 5
• identify that animals, including humans, need the right types and amount of nutrition, and that they cannot make their own food; they get nutrition from what they eat (Y3).	**3**, 5
• identify that humans and some other animals have skeletons and muscles for support, protection and movement (Y3).	5
• describe the simple functions of the basic parts of the digestive system in humans (Y4).	5
• identify the different types of teeth in humans and their simple functions (Y4).	5
• construct and interpret a variety of food chains, identifying producers, predators and prey (Y4).	**1**, **2**, 3, 4, 5
• describe the changes as humans develop to old age (Y5).	
• identify and name the main parts of the human circulatory system, and describe the functions of the heart, blood vessels and blood (Y6).	5
• recognise the impact of diet, exercise, drugs and lifestyle on the way their bodies function (Y6).	5
• describe the ways in which nutrients and water are transported within animals, including humans (Y6).	3, **5**, **12**
Everyday materials	**Relevant chapters**
Pupils should be taught to:	
• distinguish between an object and the material from which it is made (Y1).	**8**, **9**, **10**
• identify and name a variety of everyday materials, including wood, plastic, glass, metal, water and rock (Y1).	**8**, **9**, **10**, **12**
• describe the simple physical properties of a variety of everyday materials (Y1).	**8**, **9**, **10**

(Continued)

[bold] – indicate particular/extended coverage
[non-bold] – indicate the targets are threaded throughout the chapter.

(Continued)

• compare and group together a variety of everyday materials on the basis of their simple physical properties (Y1).	**8, 9, 10**
• identify and compare the suitability of a variety of everyday materials including wood, metal, plastic, glass, brick, rock, paper and cardboard for particular uses (Y2).	8, **9, 10**
• find out how the shapes of solid objects made from some materials can be changed by squashing, bending, twisting and stretching (Y2).	8, 9

Seasonal changes	Relevant chapters
Pupils should be taught to:	
• observe changes across the four seasons (Y1).	**7, 13**
• observe and describe weather associated with the seasons and how day length varies (Y1).	**7, 13**

Living things and their habitats	Relevant chapters
Pupils should be taught to:	
• explore and compare the differences between things that are living and dead, and things that have never been alive (Y2).	1, **5, 6**
• identify that most living things live in habitats to which they are suited, and describe how different habitats provide for the basic needs of different kinds of animals and plants and how they depend on each other (Y2).	1, **2**, 3, **4**
• identify and name a variety of plants and animals in their habitats, including microhabitats (Y2).	**1, 6**
• describe how animals obtain their food from plants and other animals, using the idea of a simple food chain, and identify and name different sources of food (Y2).	1, **2**, 3, 4
• recognise that living things can be grouped in a variety of ways (Y4).	1, 2, **6**
• explore and use classification keys to help group, identify and name a variety of living things in their local and wider environment (Y4).	1, 2, **6**
• recognise that environments can change and that this can sometimes pose dangers to living things (Y4).	**5**
• describe the differences in the life cycles of a mammal, an amphibian, an insect and a bird (Y5).	**5**
• describe the life process of reproduction in some plants and animals (Y5).	4, **5**, 6
• describe how living things are classified into broad groups according to common observable characteristics and based on similarities and differences, including microorganisms, plants and animals (Y6).	1, **2**, 3, 4, **6**
• give reasons for classifying plants and animals based on specific characteristics (Y6).	1, **2**, 3, 4, **6**

[bold] – indicate particular/extended coverage
[non-bold] – indicate the targets are threaded throughout the chapter.

Rocks	Relevant chapters
Pupils should be taught to:	
• compare and group together different kinds of rocks on the basis of their appearance and simple physical properties (Y3).	9, **10**
• describe in simple terms how fossils are formed when things that have lived are trapped within rock (Y3).	9, **10**
• recognise that soils are made from rocks and organic matter (Y3).	**10**

Light	Relevant chapters
Pupils should be taught to:	
• recognise that they need light in order to see things and that dark is the absence of light (Y3).	**16**
• notice that light is reflected from surfaces (Y3).	**16**
• recognise that light from the Sun can be dangerous and that there are ways to protect their eyes (Y3).	**16**
• recognise that shadows are formed when the light from a light source is blocked by an opaque object (Y3).	**16**
• find patterns in the way that the size of shadows change (Y3).	**16**
• recognise that light appears to travel in straight lines (Y6).	**16**
• use the idea that light travels in straight lines to explain that objects are seen because they give out or reflect light into the eye (Y6).	**16**
• explain that we see things because light travels from light sources to our eyes or from light sources to objects and then to our eyes (Y6).	**16**
• use the idea that light travels in straight lines to explain why shadows have the same shape as the objects that cast them (Y6).	**16**

Forces and magnets	Relevant chapters
Pupils should be taught to:	
• compare how things move on different surfaces (Y3).	**18**
• notice that some forces need contact between two objects, but magnetic forces can act at a distance (Y3).	**18**
• observe how magnets attract or repel each other and attract some materials and not others (Y3).	**18**
• compare and group together a variety of everyday materials on the basis of whether they are attracted to a magnet, and identify some magnetic materials (Y3).	**18**
• describe magnets as having two poles (Y3).	**18**, 19
• predict whether two magnets will attract or repel each other, depending on which poles are facing (Y3).	**18**
• explain that unsupported objects fall towards the Earth because of the force of gravity acting between the Earth and the falling object (Y5).	18, **19**

(Continued)

[bold] – indicate particular/extended coverage
[non-bold] – indicate the targets are threaded throughout the chapter.

(Continued)

	Relevant chapters
• identify the effects of air resistance, water resistance and friction that act between moving surfaces (Y5).	**18, 19**
• recognise that some mechanisms, including levers, pulleys and gears, allow a smaller force to have a greater effect (Y5).	18
States of matter	**Relevant chapters**
Pupils should be taught to:	
• compare and group materials together, according to whether they are solids, liquids or gases (Y4).	**8, 9**
• observe that some materials change state when they are heated or cooled, and measure or research the temperature at which this happens in degrees Celsius (°C) (Y4).	**8, 9**
• identify the part played by evaporation and condensation in the water cycle and associate the rate of evaporation with temperature (Y4).	**8, 9**
Sound	**Relevant chapters**
Pupils should be taught to:	
• identify how sounds are made, associating some of them with something vibrating (Y4).	**17**
• recognise that vibrations from sounds travel through a medium to the ear (Y4).	**17**
• find patterns between the pitch of a sound and features of the object that produced it (Y4).	**17**
• find patterns between the volume of a sound and the strength of the vibrations that produced it (Y4).	**17**
• recognise that sounds get fainter as the distance from the sound source increases (Y4).	**17**
Electricity	**Relevant chapters**
Pupils should be taught to:	
• identify common appliances that run on electricity (Y4).	**15**
• construct a simple series electrical circuit, identifying and naming its basic parts, including cells, wires, bulbs, switches and buzzers (Y4).	**15**
• identify whether or not a lamp will light in a simple series circuit, based on whether or not the lamp is part of a complete loop with a battery (Y4).	**15**
• recognise that a switch opens and closes a circuit and associate this with whether or not a lamp lights in a simple series circuit (Y4).	**15**

[bold] – indicate particular/extended coverage
[non-bold] – indicate the targets are threaded throughout the chapter.

	Relevant chapters
• recognise some common conductors and insulators, and associate metals with being good conductors (Y4).	15
• associate the brightness of a lamp or the volume of a buzzer with the number and voltage of cells used in the circuit (Y6).	15
• compare and give reasons for variations in how components function, including the brightness of bulbs, the loudness of buzzers and the on/off position of switches (Y6).	15
• use recognised symbols when representing a simple circuit in a diagram (Y6).	15

Properties and changes of materials	**Relevant chapters**
Pupils should be taught to:	
• compare and group together everyday materials on the basis of their properties, including their hardness, solubility, transparency, conductivity (electrical and thermal), and response to magnets (Y5).	**9**, 10, 18
• know that some materials will dissolve in liquid to form a solution, and describe how to recover a substance from a solution (Y5).	**9**, 10
• use knowledge of solids, liquids and gases to decide how mixtures might be separated, including through filtering, sieving and evaporating (Y5).	8, **9**, **10**
• give reasons, based on evidence from comparative and fair tests, for the particular uses of everyday materials, including metals, wood and plastic (Y5).	9, 10
• demonstrate that dissolving, mixing and changes of state are reversible changes (Y5).	10, **11**
• explain that some changes result in the formation of new materials, and that this kind of change is not usually reversible, including changes associated with burning and the action of acid on bicarbonate of soda (Y5).	8, 9, 10, **11**

Earth and space	**Relevant chapters**
Pupils should be taught to:	
• describe the movement of the Earth, and other planets, relative to the Sun in the solar system (Y5).	**13**
• describe the movement of the Moon relative to the Earth (Y5).	**13**
• describe the Sun, Earth and Moon as approximately spherical bodies (Y5).	**13**
• use the idea of the Earth's rotation to explain day and night and the apparent movement of the Sun across the sky (Y5).	**13**

(Continued)

[bold] – indicate particular/extended coverage
[non-bold] – indicate the targets are threaded throughout the chapter.

(Continued)

Evolution and inheritance	Relevant chapters
Pupils should be taught to:	
• recognise that living things have changed over time and that fossils provide information about living things that inhabited the Earth millions of years ago (Y6).	**1**, **4**, 5, **6**, 10
• recognise that living things produce offspring of the same kind, but normally offspring vary and are not identical to their parents (Y6).	3, **6**
• identify how animals and plants are adapted to suit their environment in different ways and that adaptation may lead to evolution (Y6).	**1**, **2**, **4**, 5, **6**, **7**

CURRICULUM LINKS – SCOTLAND

The following table shows how chapter coverage links to the outcomes and experiences set out in the Curriculum for Excellence (Education Scotland).

Planet Earth: Biodiversity and interdependence	Relevant chapters
• I have observed living things in the environment over time and am becoming aware of how they depend on each other (up to P1).	**1, 2**
• I have helped to grow plants and can name their basic parts. I can talk about how they grow and what I need to do to look after them (up to P1).	**4**
• I can distinguish between living and non-living things. I can sort living things into groups and explain my decisions (P2–4).	**1, 2, 5, 6**
• I can explore examples of food chains and show an appreciation of how animals and plants depend on each other for food (P2–4).	**2**
• I can help to design experiments to find out what plants need in order to grow and develop. I can observe and record my findings and from what I have learned I can grow healthy plants in school (P2–4).	**4**
• I can identify and classify examples of living things, past and present, to help me appreciate their diversity. I can relate physical and behavioural characteristics to their survival or extinction (P5–7).	**1, 2, 4**
• I can use my knowledge of the interactions and energy flow between plants and animals in ecosystems, food chains and webs. I have contributed to the design or conservation of a wildlife area (P5–7).	**2**
• Through carrying out practical activities and investigations, I can show how plants have benefited society (P5–7).	**4**
• I have collaborated in the design of an investigation into the effects of fertilisers on the growth of plants. I can express an informed view of the risks and benefits of their use (P5–7).	**1, 2, 3, 4**
Planet Earth: Energy sources and sustainability	Relevant chapters
• I have experienced, used and described a wide range of toys and common appliances. I can say 'what makes it go' and say what they do when they work (up to P1).	**14**

(Continued)

[bold] – indicate particular/extended coverage
[non-bold] – indicate the targets are threaded throughout the chapter.

(Continued)

	Relevant chapters
• I am aware of different types of energy around me and can show their importance to everyday life and my survival (P2–4).	14
• By considering examples where energy is conserved, I can identify the energy source, how it is transferred, and ways of reducing wasted energy (P5–7).	14
• Through exploring non-renewable energy sources, I can describe how they are used in Scotland today and express an informed view on the implications for their future use (P5–7).	14
• I can investigate the use and development of renewable and sustainable energy to gain an awareness of their growing importance in Scotland or beyond (P5–7).	14

Planet Earth: Processes of the planet	Relevant chapters
• By investigating how water can change from one form to another, I can relate my findings to everyday experiences (P2–4).	12
• I can apply my knowledge of how water changes state to help me understand the processes involved in the water cycle in nature over time (P5–7).	12

Planet Earth: Space	Relevant chapters
• I have experienced the wonder of looking at the vastness of the sky, and can recognise the Sun, Moon and stars and link them to daily patterns of life (up to P1).	13
• By safely observing and recording the Sun and Moon at various times, I can describe their patterns of movement and changes over time. I can relate these to the length of a day, a month and a year (P2–4).	13
• By observing or researching features of our solar system, I can use simple models to communicate my understanding of size, scale, time and relative motion within it (P5–7).	13

Forces	Relevant chapters
• Through everyday experiences and play with a variety of toys and other objects, I can recognise simple types of forces and describe their effects (up to P1).	18, 19
• By investigating forces on toys and other objects, I can predict the effect on the shape or motion of objects (P2–4).	18, 19
• By exploring the forces exerted by magnets on other magnets and magnetic materials, I can contribute to the design of a game (P2–4).	18, 19
• By investigating how friction, including air resistance, affects motion, I can suggest ways to improve efficiency in moving objects (P5–7).	18, 19

[bold] – indicate particular/extended coverage

[non-bold] – indicate the targets are threaded throughout the chapter.

	Relevant chapters
• I have collaborated in investigations to compare magnetic, electrostatic and gravitational forces and have explored their practical applications (P5–7).	**18, 19**
• By investigating floating and sinking of objects in water, I can apply my understanding of buoyancy to solve a practical challenge (P5–7).	**18, 19**

Electricity	Relevant chapters
• I know how to stay safe when using electricity. I have helped to make a display to show the importance of electricity in our daily lives (up to P1).	**15**
• I can describe an electrical circuit as a continuous loop of conducting materials. I can combine simple components in a series circuit to make a game or model (P2–4).	**15**
• I have used a range of electrical components to help to make a variety of circuits for differing purposes. I can represent my circuit using symbols and describe the transfer of energy around the circuit (P5–7).	**15**
• To begin to understand how batteries work, I can help to build simple chemical cells using readily-available materials which can be used to make an appliance work (P5–7).	**15**

Vibrations and waves	Relevant chapters
• Through play, I have explored a variety of ways of making sounds (up to P1).	**17**
• By collaborating in experiments on different ways of producing sound from vibrations, I can demonstrate how to change the pitch of the sound (P2–4).	**17**
• Through research on how animals communicate, I can explain how sound vibrations are carried by waves through air, water and other media (P5–7).	17
• By exploring reflections, the formation of shadows and the mixing of coloured lights, I can use my knowledge of the properties of light to show how it can be used in a creative way (P5–7).	**16**

Biological systems: Body systems and cells	Relevant chapters
• I am aware of my growing body and I am learning the correct names for its different parts and how they work (up to P1).	**5**
• I can identify my senses and use them to explore the world around me (up to P1).	**7**
• By researching, I can describe the position and function of the skeleton and major organs of the human body and discuss what I need to do to keep them healthy (P2–4).	**5**
• I have explored my senses and can discuss their reliability and limitations in responding to the environment (P2–4).	**7**

(Continued)

[bold] – indicate particular/extended coverage
[non-bold] – indicate the targets are threaded throughout the chapter.

(Continued)

	Relevant chapters
• I know the symptoms of some common diseases caused by germs. I can explain how they are spread and discuss how some methods of preventing and treating disease benefit society (P2–4).	2, **3**
• By investigating some body systems and potential problems which they may develop, I can make informed decisions to help me to maintain my health and wellbeing (P5–7).	**3**, 5
• I have explored the structure and function of sensory organs to develop my understanding of body actions in response to outside conditions (P5–7).	7
• I have contributed to investigations into the role of microorganisms in producing and breaking down some materials (P5–7).	1, 2, 3
• By researching, I can describe the position and function of the skeleton and major organs of the human body and discuss what I need to do to keep them healthy (P2–4).	
• I have explored my senses and can discuss their reliability and limitations in responding to the environment (P2–4).	
• I know the symptoms of some common diseases caused by germs. I can explain how they are spread and discuss how some methods of preventing and treating disease benefit society (P2–4).	
• By investigating some body systems and potential problems which they may develop, I can make informed decisions to help me to maintain my health and wellbeing (P5–7).	
• I have explored the structure and function of sensory organs to develop my understanding of body actions in response to outside conditions (P5–7).	
• I have contributed to investigations into the role of microorganisms in producing and breaking down some materials (P5–7).	

Biological systems: Inheritance	**Relevant chapters**
• I recognise that we have similarities and differences but are all unique (up to P1).	**1, 6**
• By comparing generations of families of humans, plants and animals, I can begin to understand how characteristics are inherited (P2–4).	**6**
• By investigating the lifecycles of plants and animals, I can recognise the different stages of their development (P5–7).	**1, 4, 5, 6**
• By exploring the characteristics offspring inherit when living things reproduce, I can distinguish between inherited and non-inherited characteristics (P5–7).	**6**

Materials: Properties and uses of substances	**Relevant chapters**
• Through creative play, I explore different materials and can share my reasoning for selecting materials for different purposes (up to P1).	**9**
• Through exploring properties and sources of materials, I can choose appropriate materials to solve practical challenges (P2–4).	**9**

[bold] – indicate particular/extended coverage
[non-bold] – indicate the targets are threaded throughout the chapter.

	Relevant chapters
• I can make and test predictions about solids dissolving in water and can relate my findings to the world around me (P2–4).	9
• By contributing to investigations into familiar changes in substances to produce other substances, I can describe how their characteristics have changed (P5–7).	9
• I have participated in practical activities to separate simple mixtures of substances and can relate my findings to my everyday experience (P5–7).	9
• By investigating common conditions that increase the amount of substance that will dissolve or the speed of dissolving, I can relate my findings to the world around me (P5–7).	9

Materials: Earth's materials	Relevant chapters
• Throughout all my learning, I take appropriate action to ensure conservation of materials and resources, considering the impact of my actions on the environment (P2–4).	10
• Having explored the substances that make up Earth's surface, I can compare some of their characteristics and uses (P5–7).	10

Materials: Chemical changes	Relevant chapters
• I have investigated different water samples from the environment and explored methods that can be used to clean and conserve water, and I am aware of the properties and uses of water (P5–7).	11
• I have collaborated in activities which safely demonstrate simple chemical reactions using everyday chemicals. I can show an appreciation of a chemical reaction as being a change in which different materials are made (P5–7).	11

Topical science	Relevant chapters
• I can talk about science stories to develop my understanding of science and the world around me (up to P1).	1, 2, 3, 4, 5, 6, 7, 8, 9, 10, 11, 12, 13, 14, 15, 16, 17, 18, 19
• I have contributed to discussions of current scientific news items to help develop my awareness of science (P2–4).	1, 2, 3, 4, 5, 6, 7, 8, 9, 10, 11, 12, 13, 14, 15, 16, 17, 18, 19
• Through research and discussion I have an appreciation of the contribution that individuals are making to scientific discovery and invention and the impact this has made on society (P5–7).	1, 2, 3, 4, 5, 6, 7, 8, 9, 10, 11, 12, 13, 14, 15, 16, 17, 18, 19
• I can report and comment on current scientific news items to develop my knowledge and understanding of topical science (P5–7).	1, 2, 3, 4, 5, 6, 7, 8, 9, 10, 11, 12, 13, 14, 15, 16, 17, 18, 19

[bold] – indicate particular/extended coverage
[non-bold] – indicate the targets are threaded throughout the chapter.

COMPANION WEBSiTE

EXPLAINING PRIMARY SCIENCE is supported by a range of chapter-specific **video experiments** to support your teaching, and these are available at: https://study.sagepub.com/chambersandsouter

VIDEO EXPERIMENTS
LiKE THESE

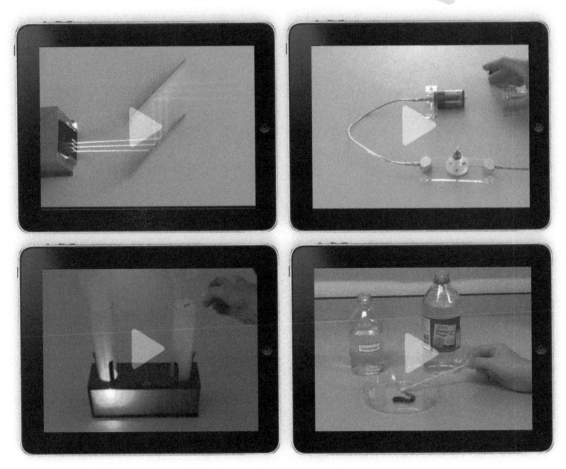

Chapter-specific **video experiments** available at:
https://study.sagepub.com/chambersandsouter

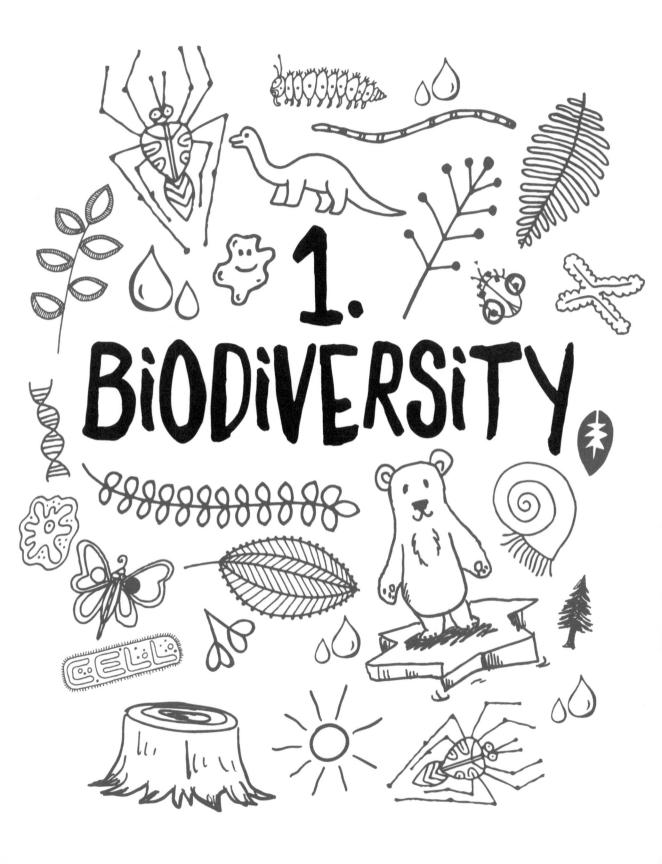

1.
BIODIVERSITY

Learning Objectives

Having read this chapter you should be able to:

* understand three big biological ideas – biodiversity, adaptation and evolution
* justify out-of-classroom learning – first-hand direct experiences of living things
* introduce observation skills
* collect and sort living things
* introduce non-flowering plants, flowering plants and animals
* practise skills – the pooter, pitfall trap, sweep net, hand lens (loupe)
* preserve plant materials – pressing, rubbing, preparing leaf skeletons
* provide examples of losses and gains in biodiversity

Overview

Biodiversity is one of the 'big ideas' in contemporary life science, and relates to all those living things that exist on Earth and as such defines not only the full range of all living things but also the underlying genetic diversity. Two other big ideas are associated with the concept of biodiversity and these are *adaptation* and *evolution*. It seems fairly straightforward to acknowledge that living things are adapted to their local surroundings, and this can be illustrated in examples such as the polar bear which is characterised by its thick white coat which provides insulation and camouflage in its environment. Adaptation is a feature of all living things and it is remarkable to consider that many engineering principles have been displayed during evolution (e.g. the structural reinforcement within trees or beeswax, the appearance of valves to ensure the one-way flow of blood, or the capacity of the retina to respond to light). It is also interesting to note that the same answer can be supplied to different problems – if we think about the camel for example, it is more difficult to explain its thick fur in terms of camouflage or keeping it warm. We therefore need to refine our thinking and consider those ideas relating to insulation and preventing *heat transfer*.

Adaptation drives the process as well as being the product of evolution: the main idea of evolutionary theory is that adaptation allows an organism to become more suited to a particular environment, and is capable of surviving environmental extremes – thus those members of a species that possess favourable adaptations are capable of surviving environmental extremes and catastrophes. These ideas were first outlined by Charles Darwin and Alfred Russel Wallace over a series of publications in the mid-nineteenth century. Wallace's contribution is frequently overlooked but cannot be underestimated.

The theory of evolution by natural selection is a pivotal idea about which Dobzhansky (1973) asserted:

'Nothing in biology makes sense except in the light of evolution'.

In Photo 1.1 this springtime plant, a member of the cabbage family, has found a new habitat on roadside verges in the UK that have been treated with road salt. It is normally found on the coast and is adapted to surviving salt spray, but has managed to move inland along motorways and roads.

Photo 1.1

© N Souter

All living things show 'descent with modification'. This means that organisms inherit characteristics from their parents who in turn inherited them from their own parents. Inheritance goes back into time.

The changes that happen in the genetic make-up of an individual can lead to it becoming more or less suited to the environment in which it lives. The genetic variation that exists within populations is brought about by sexual reproduction and mutation. Beneficial adaptations give rise to changes in the population which in turn lead to the establishment of new species.

Descent with modification is one of the main ideas in the Theory of Evolution by Natural Selection. This theory is accepted by scientists as it is supported by an overwhelming body of evidence from a wide range of fields, including anatomy, behaviour, palaeontology, geography, molecular biology, and embryology. Evolution continues to generate *public* controversy and discussion, despite Darwin's and Wallace's ideas being outlined more than a century and a half ago. There is however no real controversy in science about the Theory of Evolution. It is disputed, especially where human evolution is concerned, on the basis of beliefs by creationists and members of the 'intelligent design' movement. It seems that tensions are derived from the philosophical differences that exist between science and faith beliefs – the empirical, evidence-based, hypothesis-testing nature of science generates data that can be further tested and scrutinised by peers, while faith beliefs, by their nature, involve the acceptance of particular ideas in the absence of evidence. We do not intend to enter this debate here, rather we would accept and note that philosophical tensions remain and go on to focus on the scientific basis of the theory.

Learning and teaching

Children are intrigued from the earliest stages by the diversity that exists in familiar surroundings. This is widely encouraged through visits to wildlife parks, zoos and reserves, as well as providing a prominent position in children's literature, television and motion pictures.

The enormity of biodiversity provides conceptual difficulties. Consider, for example, the variety of birds that visit your own locale, and the difficulties you might experience in recognising, identifying and cataloguing these. The authors' home city, Glasgow, has more than 1500 species of flowering plants listed – and this is hardly an area that casual observers would regard as being biologically diverse! When we start to consider the range of mammals, birds, amphibians, insects, spiders and other invertebrates let alone the mosses, fungi and microorganisms that may be found locally, then we begin to get an idea of local biodiversity which then requires deeper consideration on a global scale.

We are fortunate in many ways to experience the outstanding standards of popular wildlife broadcasting which celebrate exotic, unusual and rare species from around the world. Yet it is also paradoxical to realise that we have poorly developed ideas of the extent of biodiversity on the planet. Statistics vary on the number of different living things that are found on the planet, depending on the source. Some scientists have estimated that there are between 3 to 30 million existing species, of which 2.5 million have been classified, including 900,000 insects, 41,000 vertebrates and 290,000 plants; the remainder are invertebrates, fungi, algae, and microorganisms.

Out-of-classroom learning

Biodiversity provides the perfect opportunity to establish ideas within a local context. We have lots of experience of celebrating biodiversity through topic studies of habitats such as Antarctica or the rainforest. However, environmental educators believe that the greatest impact that can be made on children's learning relates to raising awareness and familiarity of the living things that surround them every day. 'Think globally, act locally' has been a widely-used principal in several different contexts, and has a useful role to play in building our understanding of appropriate learning experiences for children. Out-of-classroom learning provides opportunities for cross-curricular, interdisciplinary or topic-based approaches. Our principal interest in out-of-classroom learning is that it provides the opportunities for first-hand direct experiences of living things in the immediate and consequently familiar environment. We recognise that careful planning, including alternatives during poor weather, is required to ensure that the activities are meaningful and worthwhile, that they are supported by effective briefing, and are followed up to reach a suitable conclusion upon completion. In such circumstances teachers need to take account of safety considerations, prior experiences, and learning styles.

School grounds

A piece of land or water in the school grounds, set aside and managed for wildlife, can be invaluable for encouraging the study of nature, awareness of the environment, and respect for wildlife. For teaching purposes it has the advantage of being conveniently close to school, and can be designed to demonstrate particular ecological principles such as succession. Producing such a habitat is also likely to encourage pupils to become actively interested in conservation, and can call upon the involvement of all age groups in the school.

Local resources

There is a wide variety of suitable locations which are maintained by voluntary and statutory conservation bodies and local authorities. In most cases these are primarily concerned with the protection of wildlife and habitats. All those which have experienced rangers can supply an educational service. They generally provide convenient areas where visitors can enjoy a wide range of recreational pursuits, such as water sports and fishing, in pleasant surroundings. Most have nature trails and interpretative displays, and some have visitor centres.

Visits to the seashore provide especially profitable opportunities for observing molluscs, crustacea, cnidaria, and echinoderms, while a visit to woodland and other countryside areas provides particular opportunities to observe insects, spiders, centipedes, millipedes, and earthworms.

ACTIVITY

Animal, vegetable or mineral?

The Victorian parlour game, *Animal, Vegetable or Mineral?* provides a useful starting point in the early years for grouping things according to particular properties. Prompt cards can help provide structure and encourage purposeful talking.

Teaching observation skills

A traditional model of science suggested a cycle of activities that included observations, hypotheses, experiments, results, and conclusions. Science curricula have ensured that a major element of programmes of study for science includes experimental and investigative work, as well as the constituent sub-skills. The focus has however concentrated on establishing 'systems' to ensure that learners undertake the key operational aspects of each investigation – planning, generating questions, resource preparation, experimenting, results analysis, the formation of conclusions, and evaluating the process. During investigations, scientific observations are made under carefully controlled circumstances. These are carried out to ensure that all variables – dependent, independent and control – are monitored.

Perhaps less emphasis has been placed on the nature and quality of the initial observations that pupils might make. In one study (Haslam and Gunstone, 1996) high school science students' views were examined about the processes of observation in their learning of science, as well as how they approached the task of observing during their science experiments. Amongst the findings was the disturbing point that many students regarded observation to be a teacher-directed process.

The process of making observations is a very important part of the scientific method, yet it is possibly a neglected area of science teaching. Making accurate and informed observations is a required skill for all scientists, in all fields. Initial observations, if carefully structured, can reveal much about the topic.

Careful examination of objects and events are essential to effective science. Scientific observations are, by necessity, detailed and require technical accuracy. Observation incorporates a set of skills that are of central importance in science education.

ACTIVITY

Observation – the fundamental skill

Provide a selection of objects for observation. Start with familiar things such as a pencil, and then progress to unfamiliar objects such as unusual shells, seeds, or electronic equipment. The more unfamiliar the object is, the better.

Pass a pencil around a group, asking them to add observations (i.e. those things that they may see, feel, and smell etc.). They may regard the task as low grade and start by being somewhat flippant – however as the descriptions increase in number, the task will become progressively more demanding: 'This object is hard'; 'This object is blue'; 'This object is long'... ; and so on. Do not allow them to name the object since that will destroy the purpose of the game (i.e. to make accurate descriptions). Alternatively record a list of the descriptions. How many descriptions can each class come up with?

There are some typical examples displayed in Figure 1.1.

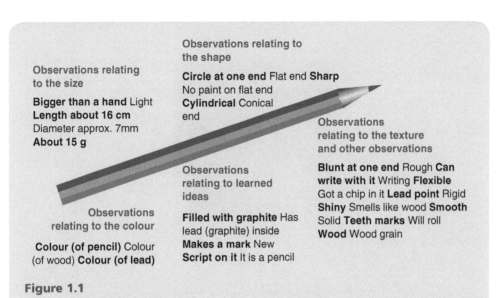

Observations relating to the size

Bigger than a hand Light **Length about 16 cm** Diameter approx. 7mm **About 15 g**

Observations relating to the shape

Circle at one end Flat end **Sharp** No paint on flat end **Cylindrical** Conical end

Observations relating to the texture and other observations

Blunt at one end Rough **Can write with it** Writing **Flexible** Got a chip in it **Lead point** Rigid **Shiny** Smells like wood **Smooth** Solid **Teeth marks** Will roll **Wood** Wood grain

Observations relating to learned ideas

Filled with graphite Has lead (graphite) inside **Makes a mark** New **Script on it** It is a pencil

Observations relating to the colour

Colour (of pencil) Colour (of wood) **Colour (of lead)**

Figure 1.1

Source: Souter & Lewthwaite, 2007

Reproduced with permission from the Association for Science Education.

Recap on the strategies deployed in making the observations

Reiterate the types of observation: those that could be measured (e.g. mass, length, diameter etc.); those that could use other mathematical, *quantitative* descriptions (e.g. shape – cylindrical, cone at end etc.); those that describe *qualitatively* such as colour, texture, taste. Then illustrate the different data types.

Once learners have secure ideas, start to formulate rules for observation about how they must describe objects in detail: they can then progress to making observations on familiar and unfamiliar natural or unnatural objects. These tend to be more difficult to describe, however this is only partly due to their irregular and unfamiliar appearance and also limitations in the use of language – but that is another conversation altogether!

ACTIVITY

Looking closely

Invite the children to make 'binoculars' with their fingers and thumbs, and then look at the lawn in front of their feet as shown by the student teachers in the photo:

(Continued)

(Continued)

Photo 1.2

© N Souter

The 'binoculars' limit the area for observation, thereby simplifying what might otherwise be complex and potentially overwhelming. This approach provides a sample area that is manageable. Initial observations which suggested they could only see 'grass' quickly change to recognition that they can see several different leaf shapes and sizes.

ACTIVITY

The famous 137-second plant hunt

Set the task for each learner to collect a single specimen of as many different leaves as possible during 137 seconds. (Rule out any windfall leaves since these may have arrived from elsewhere.) The area can be anywhere providing it has plants growing there. Invite learners to make initial observations so that the chosen area might be described as 'our lawn', 'the area that we play in', 'the sports field', the edge of the car park etc. The time is not at all important and any wacky number will suffice, although a short timescale is suggested here to provide a sense of intrigue and of urgency. The task helps to avoid the problems associated with botanical identification – the point is for each individual to find a number of their 'species' without attempting to name them – as well as impressing the group with the variety of plants that they can find in small areas in their school grounds. Our experience is that classes readily come up with between 15–18 'species' at any time of the year, but this can rise to more than 30 with enthusiastic groups during those seasons when plant abundance is at its greatest level.

Pick a sheltered area, or return indoors, to review the findings – otherwise the leaves will blow away! How many different kinds of leaf has each pupil found? How are they sure they are different? Discuss their observations in terms of shape, size, colour, smell and so on. Select a leaf that one pupil has found and lay it on the ground. Has anyone else found the same one? And are they sure? Opportunities will arise to consider the variation that exists within a population. Place all similar leaves in a pile. Give a pupil who did not find that type a specimen, and then ask them to find another one the same elsewhere in the grounds.

What sorts of conditions might favour the growth of this plant? Progress to a second leaf found by another pupil, and repeat until all the various types collected by the whole class are located in the collection. What was the greatest number gathered by one person? What was the total number of 'species' collected by the class?

Key Life Science outcomes are introduced within this activity:

- *Biodiversity*, as a concept, is established in a local context – i.e. outside the classroom door.
- *Abundance* is established through the comparative numbers of each leaf type found; abundance of plant species in turn determines the associated food chains and food webs.
- *Adaptation* of plant species and their distribution and occurrence being related to the physical environment, shade, damp areas etc. is made explicit.
- *Variation* within a species is highlighted.
- *Observation skills* are practised and need to be rigorous during collection and comparisons.
- *Identification* although tentative is based on observable differences between the selected items; individuals make decisions on the basis of similarities and differences; groups extend this practice.
- *Collecting and preserving* specimens can start a lifelong interest in natural history.
- *IDL (Interdisciplinary learning)* opportunities arise in terms of image capture, data presentation, spreadsheet use etc.

Collecting and sorting things

Background

Collecting involves using a variety of approaches for trapping and collecting specimens and then working with two linked skills – identification and classification.

Identification requires careful observation in relation to features such as shape, size, colour etc., and *classification* involves locating these observations to conventional groupings that have been agreed by scientific communities. Contemporary approaches in identification frequently rely on DNA analysis.

Anatomical, physiological, and behavioural adaptations lie at the heart of Darwin's Theory of Evolution.

Learning and teaching

The principal focus is to highlight the differences between major groups of living things without necessarily dwelling on scientific dogma. We suggest focusing on groups that are accessible and can be collected, observed, recorded, and returned to a suitable location. Identification of species involves specialist skills that lie beyond the experience of most of us. A key point is that you can start to place the organism in a suitable group – vertebrates, invertebrates, plants, fungi etc.

Organising collections of items provides a sense of order within which items can be placed into related groups and at a later date. This intrinsically human activity has applications in a range of fields including mathematics, information and communication technology, languages and the social sciences. These ideas are developed more fully in Chapter 6.

It is not always easy to determine if something is living and indeed to be convinced by this. Think for example of sponges or coral which have the appearance of being inert, inorganic materials while they are in fact significant members of marine ecosystems. It is worth considering the seven characteristics of living things – feeding, movement, breathing or respiration, excretion, growth, sensitivity, and reproduction. Children have difficulties suggesting that plants are alive but will still recognise that they can grow. The rapid responses of sensitive plants, such as the Venus flytrap, emphasise that plants are living. Defining living things is really quite complicated and children might have intuitive ideas that can be built upon.

ACTIVITY

A starter question might look at a set of items e.g. sheep grazing, a fire, a garden shrub, a wooden plank, the Sun, seeds in a packet, a potato, a river, a battery-operated toy, and ask which of them are alive?

Invite groups to browse old magazines to look at photographs or drawings and discuss which items are alive and which are not alive. Once they have agreed they should cut out no more than 10 things that are alive and 10 things that are not alive. Share their findings with one or two other groups by placing their cuttings inside two hoops on the floor that are labelled 'alive' or 'not alive'. (An alternative third classification is possible by including 'once alive' and providing a third hoop or an overlap between the first two.)

Ask the groups to explain their choices; seeking criteria for saying something is alive. The most likely qualities will include eating, breathing and growing. While this is an incomplete list it provides a useful starting point upon which future targets might be established.

We suggest that several of the ideas related to the characteristics of life are beyond the scope of early stages and can be revisited during later years. Determining that something is living by its growth and movement may be sufficient at this stage.

You might take the opportunity of organising data on index cards, spreadsheets, and databases.

'Top Trumps' games often provide useful visual materials that show differences e.g. between dog and cat breeds. Collections of single fruits or vegetables from the supermarket can illustrate a wide range of varieties within a species.

Identifying living things

Sorting things into groups

It is impossible to say how many different numbers of living things exist on Earth. This is because many of the living things are tiny and a lot of the world remains unexplored.

Roughly 1.8 million species have been properly recorded and identified. Scientists estimate that between 3 and 30 million species actually exist. They have already identified 290,000 different kinds of plants.

Identification databases

Scientists keep the information on all known things in identification databases (i.e. computer records which maintain a whole lot of information). Identification databases exist in a number of specialist areas (e.g. mammals, insects, fungi, and flowering plants). Each of these is an enormous online directory that allows scientists to identify living things or their parts (e.g. bones, wings, spores, or flowers).

Sort out that mess!

Scientists have organised all living things into different groups. For example, hawthorns are found around the northern hemisphere.

The hawthorn is a flowering plant that produces seeds. This helps us to separate it from a whole lot of different plants (e.g. mosses and ferns) and narrows our search to around 290,000!

Botanists have grouped the hawthorn in the rose family. They have done this because hawthorns have many similarities with the flowers and fruits of other members of the rose family. This narrows our search down to around 2,500 species.

The possibilities are further narrowed down as hawthorn fruit is similar to other members of the rose family (e.g. plums and peaches). There are only about 200 members in this group.

Detailed identification is made when we look at the fruit and find that it contains just one seed.

The millennium seed bank

This Royal Botanical Garden's conservation project collects and stores seeds from around the world. Seeds are cleaned, dried and identified, and then stored at a sub-zero temperature. From time to time some seeds from each collection are germinated to make sure that they're still alive, and when the collection becomes too small the seeds are grown through the plant's entire life cycle to produce new seeds.

The project aims to maintain a seed store of all known plants which helps preserve a plant if it is at risk of becoming extinct. It also helps if a particular plant has useful properties that scientists want to exploit (e.g. in the production of new foods and medicines).

Non-flowering plants

Non-flowering plants reproduce with spores or seeds. The major groups of non-flowering plants have particular distinguishing features. Algae, moss, horsetails, ferns, and conifers are all non-flowering plants. Direct observations of members of these five groups will highlight the observable as well as the ecological differences between them.

Background information

The principle of seed-bearing or non-seed-bearing plants is further emphasised as the basis of the groupings. It should be noted that significant botanical differences exist between true flowering and the cone bearing plants (angiosperms and gymnosperms): although conifers produce seeds these are open to the environment and fall out of the cones when they ripen, while the seeds of true flowering plants are enclosed within the fruit as outlined previously. These differences are subtle, but important.

Algae

Look for the presence of algae on damp walls or tree trunks in the school grounds. Algae grow on the damp, shaded, north sides of buildings and trees, and can be used to indicate direction.

Grow algae in a two-litre plastic bottle. Place one and a half litres of tap water in the bottle and leave overnight to let the disinfecting chlorine present in tap water to escape from the water. Place 15 drops of liquid plant fertiliser (e.g. Baby Bio) in the water to provide nutrients for the algae to grow. Inoculate the bottle with 30–40ml (two tablespoons) of pond water and leave in a bright, illuminated place (e.g. a window sill). Several varieties of algae will grow in a few weeks, appearing as green material, predominantly on the illuminated side of the bottle.

Moss

Mosses are prevalent in damp places and an extraordinary diversity, almost 800 species, exists in the British Isles. This is much more than the rest of Europe due, principally, to climatic conditions. They are worth studying since they're such a notable part of our vegetation. The paucity of professional scientists in this area

Photo 1.3

© N Souter

means that informed amateurs are able to develop expertise and contribute significantly to this understudied field.

Mosses are mainly green, flowerless plants that lack an internal transport system. Their leaves have no protection against water loss. Mosses reproduce sexually by dispersing spores that are released by stalked structures called sporophytes.

Moss reproduces asexually with underground structures e.g. tubers

A tremendous advantage that they have over other plant groups, in learning and teaching terms, is that they are visible and available throughout the year, in all seasons.

Initial studies can be made with simple apparatus – a hand lens and a spray bottle are enough to get you started. The hand lens allows you to look more closely at detailed structures and the spray bottle allows shrivelled specimens to be revitalised.

ACTIVITY

- Encourage learners to search in an area of turf for moss amongst the grass.
- Observe cushion mosses (often found in gutters, in roofing channels and on walls.)
- Make a photographic record of mosses in the local area using a Digiscope.
- Establish a moss garden in the classroom.

Horsetails

Direct experience of these non-flowering plants will permit learners to feel their unusual surface which includes silicon materials – and the resultant rough texture. Let them pull these apart and see how they piece together, in almost the same way as Lego pieces.

Horsetails' abrasive qualities led to their being used for scouring pots in ancient times.

Photo 1.4

© N Souter

Make a giant horsetail from sugar paper cylinders by making cylinders of progressively larger diameter recycled paper. Serrated edges will emphasise the 'Lego' effect. An alternative model can be made with 'bendy straws'.

Ferns

Collect and examine some ferns. These are common in woodland. Bracken invades rough pasture. Pressed specimens can be used by subsequent years' classes.

Photo 1.5

© N Souter

Look at the way that a fern leaf continually branches.

Compare the upper and lower surfaces, looking especially for their reproductive structures that appear as brown discs on the lower surface – these 'sori' release spores that allow the fern to spread into new areas.

Photo 1.6

© N Souter

Conifers

Children are curious about cones. Pine cones tend to open and close depending on atmospheric humidity (i.e. an open cone suggests dry weather and a closed one indicates damp weather).

Make a collection of different kinds of cones and encourage learners to note the differences and similarities between them.

Look for seeds between the scales of the cones as shown in the photo. Discuss how these might be spread from place to place (the seeds have a papery vane that helps their dispersal by the wind).

The developing Larch cone is so soft in springtime that you can cut it with a knife. The developing seeds and scales are clearly visible.

Photo 1.7

© N Souter

Collect pine seeds between March and May and grow these in some good seed compost.

ICT opportunities

Use a Digiscope to view the appearance of the different types of algae that are grown in the culture described above.

Press and preserve different types of ferns. Scanned or photographic images can provide a permanent record.

Flowering plants

The dominant land plants (i.e. the angiosperms) are defined botanically by the possession of seeds that are enclosed within a fruit which has been produced by flowering – their reproductive process. Flowering involves the production of sex cells, pollen and ovules within specialised reproductive structures. Flowering plants possess additional structures that ensure that fertilisation will take place, as well as those that will ensure that the seeds will be dispersed. Although there are only two divisions of angiosperms (i.e. monocots and dicots) they exhibit extraordinary diversity.

Further details on reproduction in flowering plants and their economic significance and utility are covered in Chapter 4.

The animal kingdom

The identification of living things relies on organising all living things into smaller and smaller groups. Traditionally all life forms have been grouped into either plant or animal groups. However classification systems have changed as scientists learn more about the evolutionary and molecular relationships of organisms and discover new species. The '5 Kingdoms' classification system is commonly used and is made up of monera, protista, fungi, plants and animals. Teachers may determine how much learner awareness of the plant and animal kingdoms provides sufficient insight at this stage.

Classification systems are hierarchical arrangements where living things are divided into progressively smaller and smaller groups. Each kingdom conforms to its own conventions, and here we use the tiger to show how this works in the animal kingdom:

Kingdom:	Animal
Phylum:	Chordates (animals with backbones)
Class:	Mammals (suckle their young)
Order:	Carnivores (meat eating)
Genus:	Panthera
Species:	*Panthera tigris*

Skills of observation and description are required for using and constructing keys, and the discipline of taxonomy relies on the availability of accurate keys. Many specialist keys are now held on the Internet. Taxonomy involves the systematic organisation of living things into categories that demonstrate relationships to each other and can indicate evolutionary trends. This is an essential area of biology since big issues regarding biodiversity and the gene pool depend on an accurate identification of living things. Learners' observations and descriptions require explanation (this can promote language development as well as scientific excellence), however children might need to be encouraged to make explicit their reasons for grouping/classifying as this may not happen automatically. Fresh and living material provides first-hand experiences and these should be used as much as possible. Drawings and photographs are useful secondary resources but cannot show all the significant details. A growing sense of wonder should be apparent at the variety of living things that surround us – as a teacher you should not feel threatened in the field if you are unable to answer learners' questions on the identity of living things. It is more desirable that they should celebrate the recognition of something new (to them at least) and use available resources to make a tentative identification, or at least show it to another pupil so that they may be able to find the same thing. Encourage the widest range of observations – learners should *look, smell and feel* living material to extend their observations and promote description. (Note however that taste is inappropriate for obvious safety reasons.)

Minibeasts – ground dwelling invertebrates

The term 'minibeast' has no biological meaning and refers simply to any small living creature. These are generally invertebrates and as such provide opportunities for children's learning about key biological principles. Learners are always very keen to collect minibeasts from their gardens or the playground, and as such their enthusiasm will overtake any of your own reservations – the minimal handling that is required will almost always be willingly carried out by them.

ACTIVITY

Collecting minibeasts

Ideally you should collect some creatures from the school grounds. Remind learners that living creatures must be treated with respect and returned unharmed to their habitats following observation. Ensure hands are washed following all activities.

There are several easy methods of collecting minibeasts.

A plastic sheet secured to the ground for a few days will encourage minibeasts to make a home underneath it – worms, woodlice, slugs etc. may all appear.

Place an old umbrella or sheet on the ground. When you shake an overhanging bush or tree the creatures will fall out onto the sheet. A remarkable collection can be gathered together in a short time. Although these are difficult to identify comparisons may be made between the populations that live in different tree or bush species. Collect the minibeasts with a pooter (see Activity following).

Figure 1.2

Crown Copyright, courtesy of the Forestry Commission (March 2014)

ACTIVITY

Using a pooter

A pooter is a small container such as a specimen tube with a sealed lid, and inlet and outlet tubes.

The entry of the outlet tube is covered with a piece of muslin that prevents the passage of solid materials, including the chosen creatures. It is used in the following way:

- Clean the mouthpiece with a sterile wipe.
- Point the nozzle towards the objects you want to collect.
- Suck sharply through the mouthpiece.
- Examine and identify the catch inside the chamber. (Hand lenses will help with this.)
- Return the catch to their habitat by removing the stopper and tipping them out of the specimen jar.
- Clean the mouthpiece with a sterile wipe.

ACTIVITY

Using a pitfall trap

Pitfall traps are useful for collecting ground-dwelling minibeasts. The technique for using a pitfall trap is illustrated in the figure.

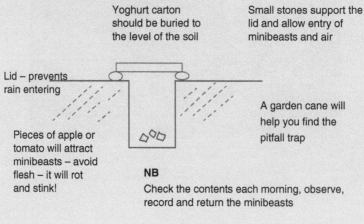

Yoghurt carton should be buried to the level of the soil

Small stones support the lid and allow entry of minibeasts and air

Lid – prevents rain entering

A garden cane will help you find the pitfall trap

Pieces of apple or tomato will attract minibeasts – avoid flesh – it will rot and stink!

NB

Check the contents each morning, observe, record and return the minibeasts

Figure 1.3

To view a video clip for these activities see the companion website at **https://study.sagepub.com/chambersandsouter**

ACTIVITY

Using a sweep net

You may be able to borrow sweep nets and pond nets from the local secondary school or ranger service who will demonstrate their application in collecting mini-beasts from water and meadows.

In meadows the technique for using such a net is to sweep it back and forward across the meadow: 10 sweeps will usually suffice – you will be amazed at how many specimens you will find. Tip these onto a tray or sheet and then capture them with a pooter for close examination.

In ponds and streams sampling can be carried out in the following way:

Check the depth of the water and that there are no obstacles that would cause children to trip.

Thirty seconds worth of sweeping is normally long enough, then empty the contents into a tray for examination.

In a pond gently sweep the net back and forward.

In a stream hold the net in a position that allows invertebrates to be carried by the current into the net.

Hand lens (loupe)

A major part of any approach to learning and teaching science in the primary school should involve making careful, structured observations in order to start making sense of the world around us. This leads onto looking more closely and, using a hand lens/ loupe as an essential skill.

A hand lens should be held close to one eye in the same place as a spectacles lens.

Focus the object by bringing it close to the lens.

Photo 1.8

© N Souter

ACTIVITY

Looking more closely

Inexpensive 'bug pots', available from local suppliers, allow learners to look more closely at their collections, and they, their teachers and the minibeasts are kept safe from one other. A magnifier on the bug pot allows closer observation.

ACTIVITY

Keeping minibeasts in the classroom

Stick insects, snails and pondlife can provide interesting and varied materials in classrooms. Pondlife is especially useful since it can be self-sustaining, provided that evaporation and contamination are avoided, and that you do not try to place too many things inside the tank. Further advice on keeping living things must conform to local guidance.

ACTIVITY

Plant materials

Collections of winter twigs and autumn leaves can provide useful reference materials.
    Windblown leaves can be collected as homework, locally and within the school grounds. They can be compared to each other, preserved collections, and to published identification charts contained within a variety of text and organisations e.g. the Woodland Trust.

ACTIVITY

Leaves

These may be collected at no risk or damage to the plants. The collection should provide an extensive range of resources upon which supplementary activities may be based.

Pressing leaves is relatively straightforward, and plenty of advice is available on the Internet. You will need clean, undamaged specimens, then lay these flat on a layer of newspaper (kitchen towel is helpful here) and apply weight over a day or two to flatten the leaves. These make attractive decoration against a window or may be made permanent or encapsulated in 250 micron laminating pouches – a heavier gauge helps with durability.

Learners should collect at least 10 different types of leaf. Alternatively the activity may be presented as a local field trip where they can make the collection in familiar and supervised local surroundings. Bring the collections together in groups or within the whole class. Leaf identification is difficult but this should not prevent further activities. It is desirable that precise identification does not get in the way of the observational processes that are being activated here.

Ask pupils to find another leaf the same as one of 'theirs'. Can they find another, by comparing with specimens from around the class? What are the key features of each leaf (i.e. shape, size and colour)? Ask them to describe each type of leaf – botanical keys use language such as 'serrated', 'lobed', and 'pinnate'. What language can children use to describe the leaf – how does it feel, look and appear? Is it shaped like a hand or does it have little leaves on the edges? Group leaves together for verification. The most notable difference between a species is that leaves are different sizes.

ACTIVITY

Leaf rubbings

For another fun autumn activity, you can make leaf rubbings. Cover a leaf with a sheet of tracing paper and rub it with a pencil, wax crayon, or piece of charcoal. You will then see its outline and texture. The rubbings can then be used for comparative purposes.

ACTIVITY

Leaf skeletons

You can often find leaf skeletons in and around leaf litter. Hard material (i.e. lignin), which provides the strength in timber, reinforces the leaf veins and makes

(Continued)

(Continued)

up the skeleton. The soft plant tissues will have been removed by decomposers (i.e. bacteria and fungi).

It is possible to prepare leaf skeletons by chemical means, and this generally involves heating the leaves in sodium carbonate (baking soda), cooling the leaf material, and carefully brushing away the softened plant tissue to leave a skeleton behind. Eye protection and appropriate attention to the hot liquid must be carefully managed.

ACTIVITY

Wild and garden flowers

Pressed flowers can also make useful collections. Pressing wild flowers is as simple as pressing leaves. (NB Take care – collecting wild flowers can influence local populations of important wild flowers. The rule is 'If you can count them, you must not collect them'.)

Losses and gains in biodiversity

Extinction

Extinction refers to the loss of the last member of a species. Examples of extinctions are well known, for example the dinosaurs, the dodo, and the passenger pigeon. Scientists will sometimes refer to local extinctions (e.g. the loss of wolves, lynx and beavers from the United Kingdom during medieval times). These are however not recognised as real extinctions as the species persists elsewhere.

Useful experiences abound in relation to studying, grouping and identifying dinosaurs, and these remain useful in terms of evolutionary timelines. Yet it is also clear that significant confusion exists (assisted to some extent by Hollywood) in relation to deep time, its sequence, and the scale of antiquity. Children experience great difficulty in relating to the recent past, and historical events during their family's lifespan, let alone in considering the deep timespan associated with cosmology and the much shorter evolutionary timescale.

The World Wildlife Fund for Nature (WWF), and other conservation organisations, express extreme concern about the increasing extinction rate for living things. While we do not know the full range of biodiversity and accept that species extinction is an inevitable part of the cycle of life, we must also acknowledge that the rate of extinction has been increasing and is attributable to human activity.

The reduction of biodiversity appears to generate comparatively low-grade responses and rarely leads to governmental action. Activists such as Greenpeace, Friends of the Earth and WWF are frequently faced with unjust hostility, despite their determined efforts to protect the habitats of wildlife across the globe. We should encourage the debate across the wider platform, and ensure that children are familiar with the overall impact that results from diminished biodiversity in species and genetic terms. Our core value of science education must be to encourage learners to evaluate issues on the basis of the evidence that is offered. Does oil extraction justify habitat destruction with the attendant loss of species? Does deforestation for cash crops mitigate against local poverty and social and financial terms? Does line fishing tuna justify the loss of dolphins?

Identification of 'new' species

Background

While we are (rightly) concerned about the increasing rate of extinction of living things, we must also acknowledge our comparative ignorance about the range and extent of those living things with whom we share the planet. Thousands of new species are discovered every year, and the average has been about 18,000 each year for the last decade.

ACTIVITY

Research new species and prepare presentations

The focus on identification at this point is the differences that can be seen between individuals – i.e. shape, size and colour. A microscope can be used to look at the detail of some organisms. DNA 'fingerprinting' technology is currently used to determine evolutionary relationships between groups of species.

Summary

Biodiversity on this planet provides an extraordinary range of living things that are the result of adaptation and evolution. Local biodiversity provides significant learning and teaching opportunities in familiar surroundings to practise fundamental scientific skills in observing, collecting, and sorting living things into meaningful groups. Rich learning experiences can be gained by children in local and distant places which will extend their insight into the variety of plant and animal life, and in particular what they can find to extend their awareness and in turn their sense of wonder at the richness and variety of living things. Experiences that are gained during formal and informal education can lead to a lifelong interest during which individuals may take action on their environment.

REFLECTION POINTS

- What range of habitats exists in your school grounds? Consider all the areas in which living things can be found (e.g. lawns, trees, shrubs and bushes). Look carefully since the most impoverished areas and urban environments will invariably provide sufficient biodiversity for the suggested activities (e.g. the plant life that colonises the area beside kerbs, fences and cracks in the paving). Walls and gutters are also often colonised by a surprising variety of plant life.
- What range of habitats exists at a short distance from your school?
- Who is able to support you locally? Consider other educational establishments, local and national government and voluntary organisations.
- Find out about each of the following:

  o The Local Biodiversity Action Plan
  o Invasive non-native species
  o The destructive impact of non-native species (e.g. *Rhododendron ponticum*, Japanese knotweed etc.)

- Investigate the conditions that are required and then grow *Mimosa pudica* in your classroom.
- Make a photographic collection that includes at least one member of each of the five non-flowering groups.
- Mammals are the class of the chordates that we belong to. Find out the characteristics of the other four chordate groups.
- Practise two of the skills that are suggested for *collecting* minibeasts.
- Practise two of the skills that are suggested for *observing* minibeasts.
- Make a collection that includes at least six different tree leaves: include their scientific name, common name, date collected and location.

2.
ECOSYSTEMS

Learning Objectives

Having read this chapter you should be able to:

- understand organisms' roles within food chains/webs
- recognise that interdependency is a major concept in environmental systems
- explain the fate of energy within food chains/webs
- emphasise the significance of microorganisms in maintaining nutrient cycles
- illustrate the predictive nature of studying food chains/webs

Overview

All living things depend on each other, so this topic contains within it a set of concepts, including feeding relationships, energy flow, recycling in nature, and progresses to include moral and ethical considerations (e.g. conservation, global warming, climate change, and animal rights). Children's understanding of the natural world should lead to a secure understanding that living things are interdependent. Animals, plants and microorganisms survive through a tangled web which allows them to interact both with each other and the environment.

Photo 2.1

© N Souter

Insects and flowers are interdependent: the moth depends on the flower, and feeds on its nectar to obtain energy; the flower depends on the moth to pollinate its flowers.

Out-of-classroom experiences should provide the default context for learning about and within ecosystems. The environment provides enrichment for children's learning.

The Outdoor Science Working Group (2011) asserted that research evidence is compelling in relation to the 'educational benefits of teaching and learning science through fieldwork in the natural and built environments', while the Field Studies Council and the British Ecological Society (in a report by Barker et al., 2002) emphasised that 'clear evidence in other areas of education that a mix of teaching and learning approaches – including "hands-on" and differentiated learning, which characterises much outdoor teaching – does help to meet the needs of the whole class'. It also helps to motivate and inspire children who may otherwise be side-lined by a more formal classroom situation (Nundy, 2001). As well as covering curriculum content, ecological fieldwork (where living animals and plants are encountered in real habitats) can help put the fun and enjoyment back into a content-dominated curriculum. Fieldwork fulfils a number of worthwhile educational objectives in an incidental way. Scientific fieldwork provides an excellent opportunity for students to work as a team, which in itself is an important part of personal and social education. It also helps them discover what it is like to work purposefully out-of-doors in varying weather conditions, learn to appreciate natural history, and link theory and observation.

The wider benefits of fieldwork can have lifelong impacts.

Learning and Teaching Scotland's (2010) justification echoed those points:

Outdoor learning experiences are often remembered for a lifetime. Integrating learning and outdoor experiences, whether through play in the immediate grounds or adventures further afield, provides relevance and depth to the curriculum in ways that are difficult to achieve indoors.

Learning outdoors can be enjoyable, creative, challenging and adventurous and helps children and young people learn by experience and grow as confident and responsible citizens who value and appreciate the spectacular landscapes, natural heritage and culture of Scotland.

The journey through education for any child … must include opportunities for a series of planned, quality outdoor learning experiences …

Fieldwork sites

School grounds

The value of local resources cannot be underestimated; significant learning opportunities can be capitalised by making observations inside the school estate and these would include making observations of living things in situ. What living things are found there? (Advice on making observations and techniques can be found in Chapter 1.) What conditions exist there that influence the presence/distribution/absence of living things? The advantages of stepping out of your classroom into the school grounds are self-evident in terms of availability, access, and health and safety considerations, and it is likely that secure learning will be derived within familiar surroundings.

Nature reserves

The term nature reserve is used somewhat loosely here to indicate the enormous variety of managed and unmanaged locations that can supply meaningful learning opportunities. Although best practice involves making effective use of local resources that deliver opportunities for learning in familiar local environments, it is also recognised that diversity of experience can be provided by visiting distant locations. Day visits and residential experiences can each provide rich learning.

Nature reserves are maintained by voluntary and statutory conservation bodies and local authorities. In most cases they are primarily concerned with the protection of wildlife and habitats, however many also offer educational opportunities and some have purpose-built educational facilities.

Parks and legislation

National parks are found all around the world. The first of these, Yellowstone National Park, was established by the US Congress in 1872. While there is no specific formula that defines national parks it is interesting to note that they range from almost 1,000,000 km² in Greenland's national park to the diminutive Penang national park (Malaysia) at 25 km². National parks generally aim to conserve natural and cultural heritage, as well as promote the sustainable use of natural resources, public understanding and enjoyment, and host communities' economic and social development. Various levels of control and management obligations exist, and alternative facilities can be found under a number of different terms, including regional parks, countryside parks, monuments etc.

Regional parks are generally extensive areas of countryside where public access and recreation are encouraged by agreement with landowners alongside traditional land uses. Country parks are relatively small areas of countryside near towns, managed for public enjoyment. These tend to be near towns or cities, making them accessible to the public. They provide convenient areas within which people can enjoy a wide range of recreational pursuits, such as water sports and fishing, in pleasant surroundings. Many have nature trails and interpretative displays and some have visitor centres. All have experienced rangers who can provide an educational service. For these reasons they offer good outdoor opportunities for learning about the environment – and most importantly *in* the environment.

Legislative control

Protected ecosystems are included in the United Nations Millennium Development Goals. The *Convention on Biological Diversity*, agreed by more than 150 countries at the 1992 Rio Earth Summit, requires member states to (Article 6):

Develop national strategies, plans or programmes for the conservation and sustainable use of biological diversity or adapt for this purpose existing strategies, plans or programmes which shall reflect, inter alia, the measures set out in this Convention relevant to the Contracting Party concerned.

The United Nations' intentions are enacted through national and local legislation. Sites of Special Scientific Interest and Special Areas of Conservation are included in the UK in line with European Parliament directives.

Natura Sites

'Natura Sites' form a network of protected areas across Europe. There are selected scientific qualities relating to protecting threatened habitats and species. These include two different categories – Special Areas of Conservation (SAC) together with Special Protection Areas (SPAs).

Sites of Special Scientific Interest

Sites of Special Scientific Interest (SSSI) are legally protected areas of land that best represent the natural heritage – and its diversity of natural features, wildlife and geology in particular.

Resources for environmental education

1. Locality and Local Organisations

It is essential that the immediate locality should be exploited to fulfil the aims of out-of-classroom learning. Visits to specialist areas such as farms or country parks should be used to enhance the local experience. Local rangers' services, libraries and specialist interest groups are keen to engage with children, and are capable of making significant contributions.

2. National Organisations

An enormous number of organisations exist which can provide educational resources, often free of charge, to schools. Prominent amongst these in the UK are the World Wide Fund for Nature (WWF), the Royal Society for Protection of Birds, and Friends of the Earth. 'Citizen science' projects provide children with opportunities to make meaningful contributions to serious studies (e.g. the RSPB's Big Garden Bird Watch or the Woodland Trust's Nature Detectives [phenology] or tree-planting projects).

3. School Grounds

Education Scotland (2011) offers the following advice:

> The local area and the wider school community provide a unique place that can be used to engage learners in understanding and investigating the richness of where they live and belong.

> Repeated visits to the same, or similar, local places and spaces allow learners to see changes and cycles daily, monthly, seasonally and annually. They also allow different perspectives on the same place, and time for children to develop questions and answer them. Similarly, using archival records such as maps, photos or engravings of the same outdoor place allows comparisons over years, decades and centuries. Encouraging young people to research the places that they visit can further develop many of the skills for learning, life and work. These skills can be also be used as a framework for later investigations into places further afield.

School grounds provide a safe location for outdoor work and free teachers from the administrative burden which is often associated with visits. As well as a place for study school grounds can prove a useful source for materials. For teaching purposes they have the advantage of being conveniently close to school, and can be designed to demonstrate particular ecological principles such as succession. A piece of land, a bird table, a raised bed, woodpile or pond set aside and managed in the school grounds for wildlife, can be invaluable for encouraging the study of nature, awareness of the environment, and respect for wildlife. Producing such habitats is also likely to encourage pupils to become actively interested in conservation, and can call upon the involvement of all age groups in the school.

4. Students' Contributions

Maintain archive copies, or outlines, of good work – it is pointless to repeat project work with each group of pupils when a theme can be developed. Incremental records can be established during an extended period that would then generate genuine data for future interrogation.

5. Equipment

A wide variety of equipment is available from educational scientific suppliers. Many resources are also capable of being constructed/assembled in school.

Trails

Planning trails

A well-designed trail can supply all the best features of a teaching programme. Trails cover a range of activities, from 'treasure hunts' to more permanent stations where predetermined activities take place. Self-pacing gives the opportunity for task completion. Graded tasks provide differentiated activities and form starting points for extended pupil enquiry/investigation. Effective teaching will always respond to those incidental learning opportunities that will inevitably occur.

Each trail will have clearly defined learning objectives upon which purposeful activities are built, while taking into account any constraints. A site's location will be equally important as well as its restrictions and permissions. A file of local trails, assembled and kept up-to-date, would be most useful for colleagues seeking trails to visit.

A preliminary visit should be carried out to evaluate the trail's suitability in terms of the time requirement, the suitability of the terrain, and the efficiency of direction markers. Will all of the pupils need to carry out every activity, or will differentiated learning be provided or followed up with peer tuition during debriefing? Remember also to evaluate the site in relation to toilets and help points in the event of an emergency.

Revise or amend any support materials in light of any insights gained during the preliminary visit, their suitability in relation to the curriculum, and the learning needs of the target group. Prepare or locate a risk assessment for the activity.

Advanced preparation will also be required to ensure that each child understands the purpose of the visit, and a suitable briefing has taken place. Each learner will need to know what is expected of them, should have mastered the prerequisite skills, and become familiar with and have access to essential equipment and apparatus, as well as understand any guidance materials.

Organising the class also means ensuring you have a suitable counting system for checking pupils and that they are suitably dressed. Make sure they are organised in pairs and groups, that they enter the trail suitably briefed, and are appropriately supervised. They must also be clear about how to respond in the event of accidents or if they find themselves separated from their group.

Finding out what's there: samples and estimates

Businesses and political parties will often sample opinions about products and policies. Likewise, when scientists want to find the number of living things (i.e. populations) in a particular area they will take a series of samples. Usually it is not possible to look at the entire environment in detail, and so scientists will sample a section or small portion that is representative of the whole area. The sampling technique used will depend on the habitat and the type of organisms present.

Samples can be taken in lots of different ways, such as by using nets and traps, and their purpose is to allow estimates to be carried out.

The sycamore leaf in the photo has several different types of insects living on it. It also has spots which appear in the autumn and are caused by a fungus. It would be impossible to count the exact numbers of leaves, bugs and spots on the entire tree. Scientists must therefore estimate the population as accurately as they can. They will

Photo 2.2

© N Souter

often estimate the relative mass of each of the populations of living things that are found in a particular place. This is called the biomass.

In all of these approaches the number of living things that you catch will depend on how many of them are living in that place, how active they are, and how far they move, as well as your skill in setting the traps and using the nets. The purpose of each approach is to make the estimate as accurate as possible (and to keep the errors as low as possible). Remember that these numbers will change from season to season and year to year.

Sampling techniques

Population analysis looks at both the density and distribution of organisms. However, counting every individual and plotting their location on a map is costly and time consuming, and such a census will apply to rare and endangered species only (e.g. Scottish wildcat and the Siberian tiger).

Sampling to provide population estimates attempts to eliminate or reduce sampling errors that would influence the statistical reliability of the results. Sampling is carried out at random to ensure that 'experimenter bias' is removed to increase reliability. Various approaches can be taken to ensure the sample is genuinely random and the probability of any individual being found is the same. (Note the continuous use of mathematical language here to ensure the significance of statistically robust experimental design.)

Estimating height

SAPS (www.saps.org.uk) point out that there 'are a variety of methods to do this using the skill processes of **estimating**, **measuring** and **calculating**'. Each of these is an important scientific skill involving numeracy that provides valid learning experiences for children.

Clinometers

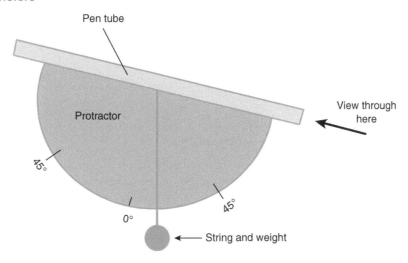

Figure 2.1

Clinometers rely on trigonometry and the simplest approach relies on the tangent of 45 = 1.

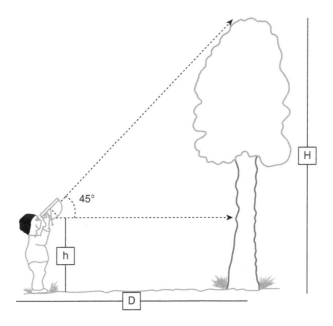

Figure 2.2

If angle A = 45° then H = D.

So to find the tree height (H + h) you must add D + h.

Tree height = D + h.

Several clinometer apps are available for tablets and smart phones, and can provide most suitable resources for use in the field.

The soil ecosystem

Soil provides an extremely rich and varied habitat. This fascinating ecosystem influences many aspects of what lives in particular areas, and supplies abundant opportunities for extensive studies of the biotic (living) and abiotic (non-living) factors that are found there.

Biotic factors in the soil

The smallest organisms – microorganisms – that are found in the soil (e.g. bacteria, protists, algae and fungi) are difficult to see without microscopes and microbiological culture techniques. The fruiting bodies of fungi, mushrooms and toadstools can provide interesting sources of biodiversity, especially during the autumn.

Photo 2.3

© N Souter

Huge numbers of invertebrates live within the soil, and these range from tiny nematode worms, springtails, insects and their larvae, to spiders and mites and a host of other organisms which contribute to soil food webs. Many of these invertebrates can be trapped by using a pitfall trap and observed more closely with hand lenses (see Chapter 1).

The signs that larger animals leave behind include pawprints, droppings and other remains (e.g. pellets, gnawed bark and burrows are commonly used to estimate their populations).

Photo 2.4

© N Souter

In the photo a squirrel has patiently removed all the scales from a pine cone to eat the seeds that were trapped between them.

ACTIVITY

Algae

Algae can be a nuisance on the surface of fish tanks, and although the numbers of algae that are found in soil water are comparatively low, it is possible to

(Continued)

(Continued)

dramatically increase the numbers. A culture can be established by shaking 5g of fresh soil in 500ml of fresh tap water inside a 200g coffee jar. Replace the lid loosely to prevent evaporation. By adding small amounts of fertiliser (e.g. a proprietary brand of houseplant food) and leaving it for a few weeks in an illuminated place such as a windowsill, the water will turn green – indicating a healthy population of algae. A class investigation could be conducted by varying the numbers of drops of plant food and recording the results photographically. Encourage learner speculation as to why this problem exists in fish tanks (i.e. fish poo acts as a fertiliser.)

ACTIVITY

Pellets

Birds form pellets from the remains of food that they cannot digest (e.g. bones, feathers and fur). They cough these pellets onto the ground while they are roosting. Different species will produce different shapes and sizes of pellets, and these provide signs that the animal is there. Learners can also dissect predators' pellets and find out what they have been eating.

Photo 2.5

© N Souter

Larger animals

Although we are surrounded by lots of wildlife it is not always easy to see it. Animal behaviour can cause this (e.g. when they're camouflaged, avoiding encounters, or nocturnal). It is therefore quite exciting when we spot creatures, even common ones, in the environment. In the UK foxes are common in towns and the countryside and yet we don't necessarily see them very often. Twenty-three species of whales, dolphins and porpoises are also common around the coastline, and yet a very small number of people have managed to see any of these.

ACTIVITY

Animal signs

1 Plaster casting can provide permanent records of animal signs, especially footprints that have been left in mud, and can be prepared with plaster of Paris. Note that the various stages require preparation, practice and patience.

Resources include learners preparing strips of card (50 × 297 mm) to construct collars around the print by securing the card with paper clips. Alternatively circular sections cut from plastic drinks bottles can be used and metallic food rings are a further alternative. Petroleum jelly should be used to line the collar and prevent the plaster of Paris sticking to it. A suitable container (e.g. a 500g margarine tub) can be used for mixing two parts plaster of Paris to one part of water with a spoon. Prior to attempting this in situ, learners will need to practise mixing water with plaster of Paris during its preparation, as well as noting the chemical changes that take place, turning the mixture from liquid to solid and with the release of heat. The plaster of Paris needs to be mixed for two to three minutes until it reaches the constituency of (unwhipped) double cream.

The collar is placed around the footprint, making sure that it penetrates the mud in order to prevent the plaster from leaking. The liquid plaster must be carefully poured inside the collar, but not directly onto the footprint since that might cause it to erode. The plaster will set in a few minutes, however the longer it is left the better. The cast must be carefully removed from the mud and from the collar. Leave this overnight and gently remove dried mud from the cast which learners can now paint or varnish, use as a mould for displays in modelling clay, or use for future reference. Once this skill has been mastered it can be utilised in a variety of contexts (e.g. for printing leaves or collecting 'evidence' for a simulated forensic investigation)

2 Visit a bird reserve and attempt to count the numbers of individual species that can be seen from the hide. Encourage learners to make observations with binoculars and telescopes.

To see video clips of animal signs and footprints go to the companion website for the book at **https://study.sagepub.com/chambersandsouter**

Abiotic factors in the soil

These factors include soil texture, soil air, soil temperature, soil water, soil solution and pH, together with soil organisms and decaying matter.

ACTIVITY

Soils

Collect different soils and examine their texture. This can be done as follows:

1 Take a small amount of soil into the palm of the hand. Moisten with water.
2 Attempt to roll into a small ball.
3 If it rolls into a small ball, attempt to roll it out into a cylinder.
4 If the soil can be rolled into a cylinder, try and bend it into a ring.

   a. If it will not form a ball (a definite single grain structure)   SAND
   b. If it forms a ball but not a cylinder   LOAMY SAND
   c. Cylinder formed with difficulty but breaks when bent   SANDY LOAM
   d. Cylinder formed and bent into a ring; not sticky   CLAY LOAM
   e. Cylinder formed and bent into a ring; sticky   CLAY (If the soil has a silky feel this classification is modified to SILT)

Identification and biodiversity

Needless to say the identification of living things is a skill and one that requires specialist training – an essential skill that is in short supply and vital for understanding biodiversity (e.g. as a resource and for drug discovery), as well as for gaining insights into the 'gene pool'.

Identification

Resources that include images of living materials (e.g. leaves, winter twigs, wildflowers, butterflies and aquatic minibeasts) are widely available, and can be helpful for learners in making an initial, tentative identification.

These resources are best included when a teacher is using them with confidence in the accuracy of their identification. Frustration, for both pupils and teacher alike, arises when the specimen in question does not appear, or is at variance with the model specimen that has been presented in the resource.

Best practice thus involves comparing resource items with the materials that have been collected – it is better to start with the perfect facsimile of the organism concerned and then to ask learners to look for specimens rather than using the chart as the basis for identification. An 'I-SPY' approach can be effective here (I-SPY books are spotters' guides written for British children, and have proven particularly successful in the 1950s and 1960s in their original form, and again when relaunched by Michelin in 2009 after a seven-year gap in publishing). For many children this will represent a starting point in a lifetime of studying biodiversity. Lots of survey volunteers are sought in a range of areas requiring differing levels of expertise (e.g. as 'Nature Detectives' and for phenology or the annual RSPB birdtable survey).

Surveying plant life

Plant populations are invariably available in the school grounds and in accessible locations. In Chapter 1 we looked at techniques for encouraging accurate observations. Greater sense can be made of the distribution of plant life by collecting quantitative data using straightforward surveying techniques. Such data support the idea that various plants are unevenly distributed throughout the environment – for example, navigation can be supported when learners are told that mosses grow on the north side of trees.

Photo 2.6

© N Souter

The plantain shown in the photo can be readily identified in August and September. Their 'rosette' of leaves spreads over the ground and can be found at the edge of paths where grass has been damaged by trampling. Their distribution and abundance can be estimated by surveying a few metres away from those paths.

Quadrats and belt transects

In Chapter 1 we looked at various ways of looking closely and established that restricting the sampling area helped to diminish the overwhelming numbers of things on view. Estimating plant populations relies on taking samples, and sampling areas can be defined in a number of ways by randomly placing a quadrat within a habitat. A quadrat is used to select a particular area in order to quantify samples. The most commonly used quadrat in secondary biology classrooms is one measuring $0.25m^2$.

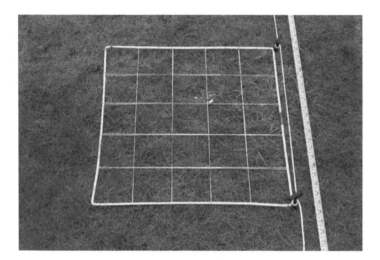

Photo 2.7

© N Souter

Note however that any regular object can be used (e.g. wire coat hangers, hula-hoops, rubber quoits). Quadrats are used to sample an area and estimate the population of organisms living there. It is a versatile biological and geographical technique, and some quadrats can be rather grand in scale – such as those that monitor the population trends of the Mexican spotted owl in Arizona and New Mexico, where randomly located quadrats of $50km^2$ are utilised, or the $100km^2$ ones that can be found in satellite-based tree surveys in Nepal. Similarly, quadrats of $10\times10cm$ can be used to estimate the population of *Pleurococcus*, a green algae that lives in the shade on tree trunks. These are used to survey plants and sessile animals such as limpets and barnacles on the seashore.

ACTIVITY

Quadrat

There are various methods of estimating plant abundance in populations using a quadrat. Abundance (i.e. how rare or common it is) can be estimated using a rubber quoit.

It is easy to see if the plant being surveyed is present inside the quoit when the flower is present, and with a little bit more attention to detail their leaves can be used for identification. Learners can repeat the sampling as many as 10 or 25 times, depending on age and stage. They should make a chart to record the number of quoits which include the plant and the numbers which don't. Note that it is important that the sampling instrument, whether it is a quoit or a quadrat frame, is placed randomly on the study area. Placing it deliberately will provide false results.

Sampling invertebrates

Various techniques for showing the presence of invertebrates have been described in Chapter 1. These can also be used to survey the invertebrate populations that are associated with different host plants.

In each case the collected specimens can be emptied onto a white plastic tray, and smaller ones can be collected for closer examination with a pooter and a hand lens (see Chapter 1). Closer examination can be carried out inside 'bug pots'.

Pooters consist of a small container such as a specimen tube with a cork, and inlet and outlet tubes. The entry of the outlet tube is covered with a piece of muslin or wire gauze (see Figure 2.3). The end of the inlet tube is placed over the animal to be collected and the operator sucks sharply through the mouthpiece. The catch is collected in the tube and can then be examined more closely.

Typical invertebrate samples include herbivorous insects and their larvae, molluscs, including slugs and snails, as well as predators including spiders. In this way comparisons of the invertebrates associated with specific trees and shrubs can be made. The abundance and diversity of populations rarely fail to impress. Identification should capitalise on children's observations (e.g. 'brown and yellow stripes', 'tiny red ones') while drawing attention to significant biological features (e.g. the number of legs, presence of wings, shape, size and colour). Further identification is complicated and may need specialist support.

Remember to return all specimens to their original location after examination.

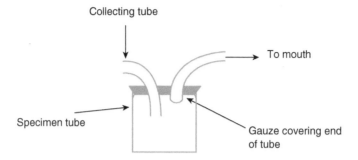

Figure 2.3

Surveying tree and shrub invertebrates

Trees and shrubs sit at the base of food webs that harbour many invertebrates. To collect these, a large white sheet or upturned old umbrella is held under the branch being studied, and the branch is vigorously shaken. Care must be taken not to damage the branch.

Surveying ground-dwelling invertebrates with sweep nets

Sweep nets are designed with a nylon or muslin bag, supported on a tough frame, and attached to a handle. Select a suitable area following a few dry days, otherwise the catch will become trapped and damaged in the wet net. Sweep the net lightly through the surface of the vegetation, trapping resting insects and other animals. The extent and number of sweeps should be standardised to ensure uniformity of sampling. It is worth remembering that the distribution of animals after dark is likely to be different from that by day.

Photo 2.8

© N Souter

The collected specimens can be emptied onto a white plastic tray or transferred with a pooter into a 'live cell' for closer examination. Return the specimens to the original location.

Surveying ground-dwelling invertebrates with pitfall traps

Pitfall traps (see Chapter 1) are used to collect insects and other small walking or crawling animals. A small container such as a yoghurt pot or plastic cup is set into a hole in the ground, so that its rim is level with the surface, but so arranged that ground water does not run down into the container. The mouth of the container should be covered to prevent rain entry.

The position of the traps must be clearly marked so that observers can find them easily. They must be emptied frequently, at least once every 12 hours, otherwise the carnivores will eat the other animals. The use of plastic cups in pairs, one inside the other, greatly facilitates emptying. Traps can be baited to catch particular insects. They can also be placed in different locations for comparisons between the populations that are associated with different trees, shrubs, or areas.

The collected specimens can be emptied onto a white plastic tray or into a 'live cell' for close examination with a hand lens. Return the specimens to their original location.

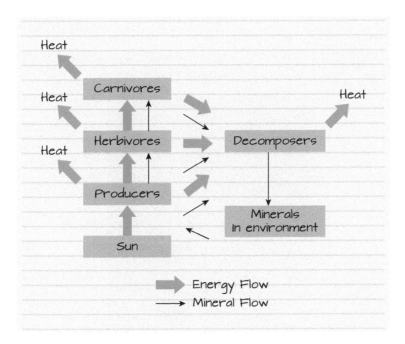

Figure 2.4

Classifying producers, consumers, predators and prey

Food chains are simple?

Food chains might be considered as being simple because they are straightforward to draw and all of them follow the same basic pattern to illustrate how living things depend on each other.

However, while food chains *might* be shown as simple illustrations, they summarise several complicated processes and contain a lot of information about feeding relationships, energy flow, and recycling in nature.

Food chains include their own language and embed a set of scientific understandings.

Producers

- Producers trap energy for their growth and then provide food for the rest of the web of life.
- Food chains need a source of energy and this can come from the Sun (photosynthesis).
- Energy can also be obtained from simple chemical sources. The microbes that grow in hot volcanic springs are able to grow and survive on the energy that they obtain from chemical reactions.

Herbivores

- The first consumers, 'primary consumers', in food chains are herbivores.
- They obtain their energy by consuming plant material.

Photo 2.9

© N Souter

Carnivores

- Carnivores eat flesh and can feed on herbivores or other carnivores.
- They are called 'secondary consumers' when they are on the second link of a chain, or 'tertiary consumers' when they are on the third link.

Omnivores

- Omnivores are more cosmopolitan in their pattern of feeding since they are able to feed on plant or animal material.
- They may be primary or secondary consumers.

Decomposers

- Decomposers break down the organic materials from dead living things or their waste materials and return the mineral content to the environment.
- Without this recycling stage producers would be unable to obtain essential minerals for their growth and the web of life could not continue.

Food webs 'simply' describe all the overlapping food chains that exist in a habitat.

Health and safety issues

Local guidelines apply to all out-of-school activities. It is essential that these are adhered to during all the preparation stages, as well as during an excursion.

Health and safety considerations should take into account the fact that significant incidents are exceptionally rare, and should not be used to minimise or avoid invaluable out-of-classroom experiences. Fisher, 2001; House of Commons, 2002 in Barker et al. 2002 noted:

> Teachers who organise out-of-classroom activities enrich the education of young people immeasurably and yet their contribution often goes unnoticed until something goes wrong. And yet if there were no such salt-of-the-earth teachers our education system would be greatly impoverished. The recent Health and Safety issues must make many dedicated teachers feel that the risk is not worth it and many a supportive school head teacher will feel the same. Fear of accidents is not confined to fieldwork; laboratory science has now become so 'risk free' that it has become bland, pedestrian and dull …
>
> (2002: 10)

Care, conduct, and conservation

'Take nothing but memories, leave nothing but footprints!' *Chief Seattle*

Successful out-of-classroom learning, as with all effective teaching, depends on careful and detailed planning. Good practice emphasises care for individuals and the environmental resources. Individual conduct should be planned in such a way as to develop sensitivity towards the environment and each other. The introduction and maintenance of conduct which conserves outdoor areas will be of benefit to all who use such areas. Contingencies related to emergencies must be included in risk assessments.

Choose and use the selected location carefully – together with accessibility these are just as significant as the activities. Areas of special scientific interest are not necessarily the best choice for beginners, and school grounds can usually provide opportunities for the widest range of experiences.

While leading activities it is important to respect ownership, be considerate towards other users, and avoid disturbing plants and animals and making unnecessary collections. Above all, remember to leave the area as you found it and give no one cause to regret your visit.

Habitats on our doorstep

All living things depend on each other and the conditions in the habitat in which they are found. Enormous differences are apparent in terms of the habitats that we can find both locally and globally. The landscape is dominated by a number of features, including the different types of rocks and weather conditions, changing day lengths between the seasons, and the way it has been managed by people.

Abiotic factors in a habitat

Each habitat includes a number of different conditions. The living things that are found there are adapted to those conditions. These physical conditions are called abiotic factors.

ACTIVITY

Abiotic factors

You might be able to use suitable equipment to compare the abiotic factors that exist in two separate habitats.

Biotic factors and survival

Living things depend on each other within the web of life. Competing for food and avoiding disease or being eaten are each important to the survival of living things on Earth.

Food is a biotic factor

The survival of a population of animals in an ecosystem depends on sufficient food being present for them.

One strategy that is carried out to ensure the survival of many groups of animals involves migration (i.e. the movement of populations from one place to another). Examples of this include wildebeest in East Africa, Pacific salmon, and many bird species. Migration is frequently driven by a need to find food and can be triggered by physical changes in the environment (e.g. temperature). Migrants find new food and can avoid extreme conditions (e.g. freezing or drought). Migration also takes place in many species to permit mating.

ACTIVITY

Bird populations

Compare bird populations at a feeding table during different seasons.

Predation is a biotic factor

While it is clear that food webs involve predator prey relationships, it will also be apparent that each population will be constantly changing. The red deer is Scotland's largest land mammal but its population needs to be controlled by 'culling' since natural predators no longer exist there. Wolves and lynx (deer predators) used to live in Scotland and regulate deer numbers naturally within the web of life.

Disease is a biotic factor

Disease, caused by viruses, bacteria, other microorganisms and parasites, leads to animals lacking energy, and being unable to feed or compete, reproduce and survive. Predators will often prey on individuals weakened by disease.

Competition is a biotic factor

Living things need resources to survive which they get from the environment. Plants compete for the physical factors that they need to grow – light, water, nutrients and space.

When poppies colonise wheat fields different competitions are taking place. Individual wheat and poppy plants are competing against another species for environmental resources – this is termed *interspecific* competition. Each individual poppy plant and wheat plant is also competing with its own species for those resources – *intraspecific* competition.

Photo 2.10

© N Souter

Why it's important to know what's there

The environment is constantly changing. Physical factors, abiotic ones, brought about by changing seasons and weather (e.g. day length, temperature, and rainfall), and geological ones (e.g. bedrock, earthquakes and volcanoes), change the conditions within habitats. Abiotic factors influence the survival of food webs within each habitat.

Changing populations

Living things (biotic factors) are also changing all the time. Populations are dynamic. Food webs are under attack from the removal of native species, the addition of new ones from elsewhere, and actions by humans (e.g. pollution and climate change).

Photo 2.11

© N Souter

Colorado beetle (*Leptinotarsa decemlineata*) larvae feed on potato plants and are a serious pest both in the United States and Europe. They are not established in the UK and are a notifiable quarantine pest, whose introduction is prohibited under the European arrangements for plant health. Their presence must be reported to the police since their population could explode and destroy UK potato crops.

Migration brings different populations to habitats at different times. Bison (the Plains Buffalo) were hunted almost to extinction during the late nineteenth century. Huge herds, with as many as four million individuals, once occupied huge areas of the North American plains. They migrated north in the spring and south in the autumn, following established trails that were as long as 500 miles.

Rabbits were introduced to Australia as soon as European sailors arrived there. Caged rabbits were grown for fresh food on their sailing ships and released into the Australian environment in the 1850s. The conditions for rabbits were ideal: a suitable climate, abundant food, lack of competition from native marsupials and absence of predators led to a population explosion. By the 1920s 10 billion rabbits were causing significant environmental destruction. The removal of plant cover contributed to soil erosion. Rabbit grazing was responsible for the extinction of many native Australian plants. A number of approaches have been taken to attempt to reduce the problem, and these have included poisoning, trapping, shooting, and the introduction of diseases.

The Himalayan balsam (*Impatiens glandulifera*) is a weed which was introduced to UK streams and river banks in 1839. It has pods which can produce up to 800 seeds. Ripe pods explode and can eject seeds to a distance of up to 7m. These are then dispersed further by the water flow and the plant can therefore spread itself rapidly. It grows vigorously, competes strongly against native plants, has few predators, and is able to thrive. It has been estimated that its presence may reduce local species diversity by as much as a quarter.

Noting the difference

The changing patterns and populations of organisms tell us much about the health of an environment. Scientists and interested members of the public record seasonal events. In springtime they might record bud bursting in oak, the appearance of the lesser celandine, nest building in rooks, or the date when they first saw a red admiral butterfly. In autumn they might record the arrival of migratory redwings, and the departure of swallows, the date they last cut the grass, or the appearance of the fly agaric toadstool (see Photo 2.3).

Phenology is the study of the times of recurring natural phenomena. Data have been collected on various seasonal events for more than a century, and records show that springtime is taking place earlier and autumn is happening later. Data provide evidence that supports the hypothesis that climate change is taking place.

Photo 2.12

© P Holmes/WTML

Many 'citizen science' opportunities, ranging from natural history to astronomy, and even the search for extraterrestrial intelligence, are available, and it is worthwhile to encourage children's participation.

ACTIVITY

Phenology

1   You might explore more about seasonal events. Encourage participation and seasonal observations using Nature Detectives.
2   Encourage children to take part using the annual RSPB's Big Garden Birdwatch.

The impact of species addition or removal

We have seen that the introduction of rabbits and Himalayan balsam or the removal of predators has resulted in habitat destruction, a reduction in biodiversity, and the extinction of many species.

Predicting impact

We can predict what might happen within food webs, but only if we understand their detail in local habitats.

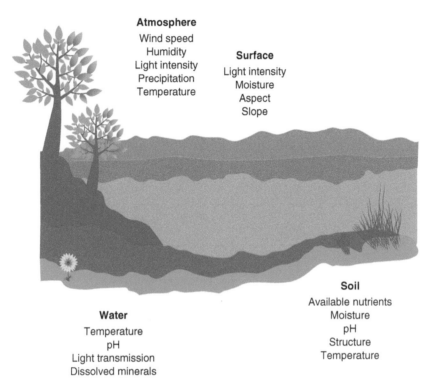

Atmosphere
Wind speed
Humidity
Light intensity
Precipitation
Temperature

Surface
Light intensity
Moisture
Aspect
Slope

Soil
Available nutrients
Moisture
pH
Structure
Temperature

Water
Temperature
pH
Light transmission
Dissolved minerals

Figure 2.5

ACTIVITY

The food web

We can speculate what the impact might be in a full range of environmental events, and it is important to recognise that influencing one member of the food web will have an effect everywhere else. Invite speculation from the children on several scenarios:

(Continued)

(Continued)

1   If the foxes were all killed what would happen to the populations of blackbirds, rabbits and mice? (They might be able to flourish since these three consumers eat different things. Their increased numbers would however place pressure on their food supply.)

2   What would the consequences be if insecticides removed ladybirds and green-fly? (The blackbirds' food source would be affected, their populations would be under pressure, and they would be working harder/competing to find earth-worms. Encourage working through the entire food web.)

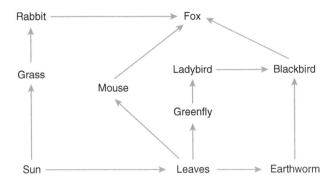

Figure 2.6

Summary

The study of ecosystems provides children with an opportunity to consider the relationships that exist between living things in their environment. It extends the concepts contained while celebrating biodiversity by providing a bigger picture that includes energy flow, interdependency, and the cyclic nature of natural systems. Ecosystems encourage a wider study of environments in order to consider both the biological and physical properties that define particular habitats, and explain the challenges experienced by living things, while also starting to explain the way in which adaptations have ensured their survival. Rather than being 'soft science', environmental assessments into population change generate data that demand the most sophisticated numerical analysis by statisticians.

REFLECTION POINTS

- Think for a few moments about an out-of-classroom learning experience that you had during your primary school years.

  Where was it? Who was the teacher? What did you do? Who did you travel with? What did you eat? You might repeat this for an event that you have managed as a teacher, and then compare these events to routine classroom practices. Which resources can be utilised for out-of-class experiences?

  - Within school grounds
  - Walking distance from the classroom
  - Locally for full day visits
  - As residential experiences

- There are lots of acronyms in this chapter (SAC, SPAs, SSSI), and there are significant legislative aspects derived from international treaties and statutory requirements, and thus a complex picture emerges. Nevertheless, these structures impact on us all and in terms of the natural environment on sustainability and development.

  How do these affect you?

  What sites of interest, conservation or protection are found locally?

  What are Local Biodiversity Action Plans?

- Consider the importance of making random samples to cross-curricular numeracy.
- Consider again the numeracy skills involved in undertaking a census, surveying, estimating and measuring.
- Consider also the ways in which pupil's record data as well as the ways that they might present the generated data.

  - What detail might be required in data analysis?
  - What resources might be incorporated into reporting?
  - What follow-up work is required?
  - How might the experiences be developed in the future – both yours and pupils?

- What kind of habitats can you see from your school or your home? What *land* habitats are there? Are you in the city or on the coast? Can you see woodland, parkland or farmland? Can you see mountains, moorland or bogs? What *freshwater* habitats are there around your school – lakes, ponds, marshes, rivers, streams or lochs? What *marine* habitats can you see? Can you see an estuary, the intertidal zone, open ocean, or rockpools?

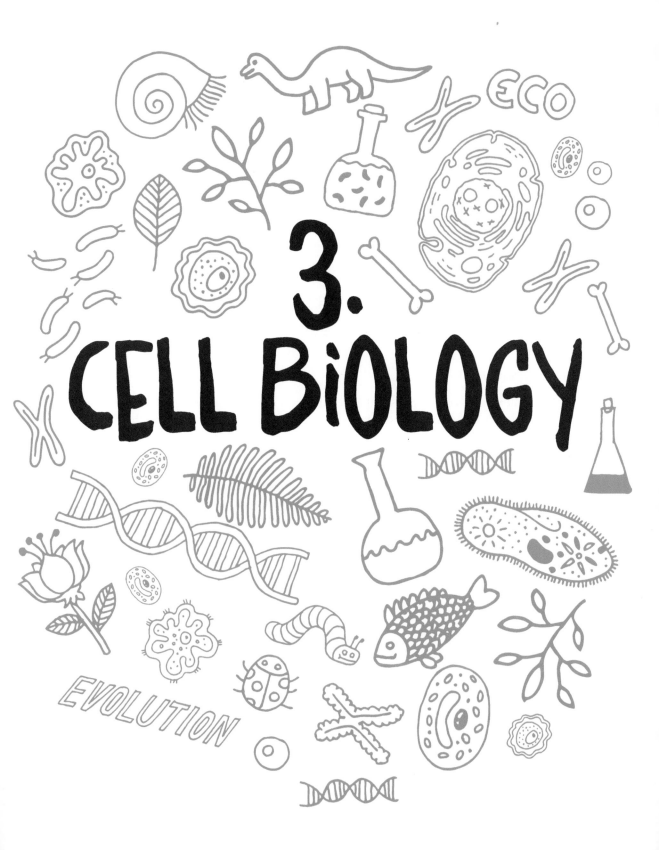

3.
CELL BiOLOGY

Learning Objectives

Having read this chapter you should be able to:

- develop insights into living processes on the basis of cell theory, that all living things are made from cells
- explore aspects related to the chemical composition of living things
- experiment to show growth and chemical changes
- observe processes in living things e.g. growth, reproduction
- carry out investigations in major biotechnological processes – baking and fermentation.
- describe the scientific and economic significance of microorganisms in agriculture and pharmaceutical industries

Overview

The context of this chapter lies within cell theory – all living things are made from cells. Essential cellular processes and aspects of cell chemistry lead to their growth and division, reproduction and repair, osmosis and diffusion (uptake and removal), industrial processes and biotechnological processing, brewing and baking, and in agriculture and pharmaceutical industries.

The significance and centrality of cell theory to science and science education have been emphasised within the seminal paper by Harlen (ed. 2010) and two of the 'Big Ideas' focus on aspects of cell theory.

Cell theory has been a central idea in life science since the 1850s.

Big Idea

Organisms are organised on a cellular basis. All organisms are constituted of one or more cells. Multi-cellular organisms have cells that are differentiated according to their function. All the basic functions of life are the result of what happens inside the cells which make up an organism. Growth is the result of multiple cell divisions.

Scientific theories

It is worth pausing to contrast what is meant by theories in science with common usage. Theories in science are widely accepted by scientists since they have been substantiated through repeated observations and experiments. Examples include 'The Big Bang Theory', 'Atomic Theory', 'Chaos Theory', and 'Evolutionary Theory'. Such theories allow scientists to make explanations of observed phenomena as well as predictions about experimental outcomes. On the other hand common usage suggests that theories involve speculation and guessing. This dichotomy may lie at the seat of some conceptual difficulties.

Scientific theories do create dilemmas for teachers since explanations frequently lie outwith concrete experiences and to children's developmental stage. For example, while we can detect the impact on bulb brightness when we add new batteries to a torch, we are unable to quantify this in terms of the voltage that is contained within the circuit. Such data need to be collected by a voltmeter. Atomic theory lies behind our understanding of chemical reactions, yet we are unable to see atoms, their interactions and constituents without sophisticated technology and resources.

We can accept cell theory, but it is only when we observe living materials more closely with a microscope that these discrete structures become apparent.

Big Idea

Genetic information is passed down from one generation of organisms to another. Genetic information in a cell is held in the chemical DNA in the form of a four-letter code. Genes determine the development and structure of organisms. In asexual reproduction all the genes in the offspring come from one parent. In sexual reproduction half of the genes come from each parent.

Cell theory – all living things are made from cells

All vital processes (e.g. growth, reproduction, variation, inheritance, energy flow, metabolism and biochemical changes) occur within living things.

Media representations make repeated references to cells in medical, agricultural and natural historical contexts. Significant public awareness exists with regard to the importance of DNA, the chemical source of inheritance, which is found specifically in the chromosome and is passed on from cell to cell during growth, as well as being passed on from parents during reproduction.

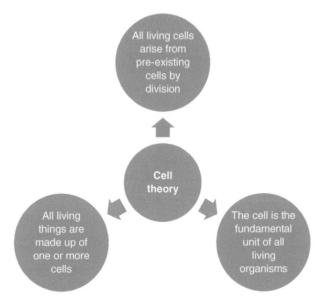

Figure 3.1

The appearance of cells

Although the cells of living things provide a bewildering array in terms of their appearance under the microscope, they fall into two basic categories: prokaryotes (pro – carry – oats) and eukaryotes (you – carry – oats). The biological distinctions, which include size and internal organisational differences, are significant, and it is important to note that the prokaryotes include microorganisms (e.g. bacteria) and the eukaryotes include plants, animals and fungi.

Prokaryotes

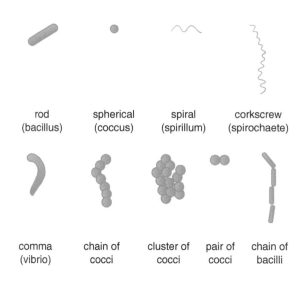

Figure 3.2

Bacterial species appear in many different shapes – e.g. spheres (cocci), rods and spirals – and in different arrangements – e.g. singly, in pairs, chains and clusters – when viewed under a high power microscope (1000 X).

Eukaryotes

Although their appearance possesses infinite variety under the microscope, eukaryotic cells are all made from the same basic structures, and organelles.

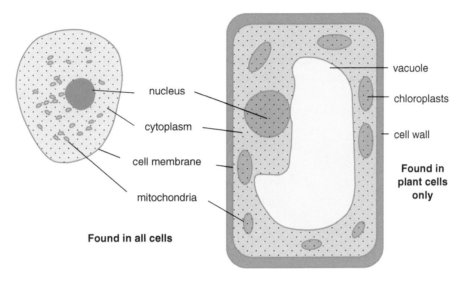

nucleus

cytoplasm

cell membrane

mitochondria

Found in all cells

vacuole

chloroplasts

cell wall

**Found in
plant cells
only**

Figure 3.3

Sex cells

Sex cells (gametes) are specialised to bring about sexual reproduction and ultimately the exchange of genetic material. These involve donor cells (male) which pass genetic material onto recipient cells (female).

Sexual reproduction takes place at the simplest level in bacteria which simply join together so that their genetic material is shared. Many different strategies are carried out in nature to ensure that the exchange of genetic material is successful. Gametes may be motile (able to swim) or sessile (incapable of locomotion). It may be surprising to learn that some plants' (e.g. mosses) male gamete is a sperm, as well as noting that the male gamete in flowering plants is incapable of moving itself and adopts an external agent (e.g. wind, water, insect).

The implications are that sexual reproduction lies at the heart of the study of inheritance. (See Chapters 4 and 6.)

Looking more closely

Scientific equipment is designed to collect data and record observations and monitor phenomena that are frequently beyond the scope of our senses (e.g. a voltmeter for voltage, a thermometer for temperature, an astronomical telescope to look at distant galaxies, and a microscope to look at cellular structures).

The digital revolution

Recent microelectronic developments have transformed our ability to look more closely at objects, capture images and videos, and provide opportunities for children to analyse and share those experiences. Inexpensive digital microscopes used to take Photo 3.1 of a shirt usually have the capability to magnify 20 to 200 times. The objects are illuminated by LEDs and connect to computers via a USB connection.

Photo 3.1

© N Souter

The advantage of image capture of this type gives teachers and learners the opportunity to meaningfully discuss the appearance of the objects in question, since they are clearly viewed and shared on screen. This contrasts starkly with the frustration encountered by many while taking turns to look through the eyepiece of a traditional light microscope, which promoted an observational guessing game of 'can you see what I see?'. JPEG files can subsequently be labelled, annotated and imported into reports and classroom displays.

Cell chemistry

All cells include the same basic structures and have the same basic chemical composition. Six elements (carbon, hydrogen, oxygen, nitrogen, phosphorus and sulphur) dominate about 98% of the mass of all living things. These combine to make a myriad of substances between which thousands of chemical reactions take place, but are coordinated to perform all the functions necessary for life. Four major kinds of large chemical compounds (macromolecules) – carbohydrates, lipids, proteins, and nucleic acids – are found within cells. Each cell's activities involve constantly organising and reorganising the chemicals that are found there in order to manage energy. Organisation within cells is carried out by particular types of proteins: their shape is maintained by structural proteins, and their activity is controlled by enzymes that are the essence of life itself.

The chemical processes inside cells (metabolism) are dynamic. Obtaining raw materials, storing and releasing energy, and waste removal require effective monitoring as well as efficient processing. Metabolism is controlled by enzymes – proteins that control reaction rates in cells.

The processes of life (growth, reproduction, waste removal and recycling in nature) all depend on energy. Almost all of the energy in ecosystems comes from photosynthesis.

Releasing energy – respiration

Respiration releases energy from food and this can take place in the absence of oxygen (anaerobic respiration) or in its presence (aerobic respiration).

Anaerobic respiration releases small amounts of energy from food materials, and forms products that still contain energy (e.g. ethanol or lactic acid), as well as releasing carbon dioxide gas. Many microorganisms are able to survive by obtaining their energy from anaerobic respiration only (e.g. baker's yeast, bacteria that colonise the digestive system, decomposers, and several pathogens).

ACTIVITY

Anaerobic respiration

Place some yeast and sugar solution in a pop bottle sealed with a balloon. Leave it in a warm place.

Aerobic respiration, which depends on oxygen, releases much higher levels of energy and completes the breakdown of the food to carbon dioxide and water. The microorganisms that rely on aerobic respiration are restricted to locations that maintain constant contact with the air. Multicellular organisms require efficient structures to ensure that oxygen uptake from the environment is maintained. Such gas exchange surfaces (e.g. lungs in mammals, gills in fish, and skin in earthworms) tend to have a large surface area for absorption, have moist surfaces into which oxygen can dissolve, and are in contact with body fluids for oxygen circulation.

Obtaining energy – photosynthesis

Almost all the chemical compounds that are found in a plant are made from water and carbon dioxide which are available to the plant from the soil and the air. Photosynthesis takes place in microorganisms and green plants.

Water and carbon dioxide contain only three elements – carbon, hydrogen, and oxygen – and photosynthesis converts these into all the organic materials necessary for their survival. In addition they obtain tiny quantities of other elements that they require to support their metabolism. All plants need elements, including nitrogen to make proteins, magnesium to make chlorophyll, potassium for protein synthesis, and phosphorous for membranes and DNA. These are essential for a plant to make the compounds it needs. Without these minerals the plant would be unhealthy and could not grow in a particular area. Crops depend on the correct minerals being present for their growth. This lies at the basis of the agricultural and horticultural use of fertilisers.

Carbon cycle

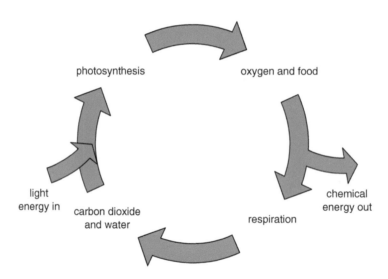

Figure 3.4

The carbon cycle illustrates how photosynthesis, respiration and energy are connected.

Food labelling

Food labelling is controlled by legislation. In the UK these labels must be clear and easy to read, permanent, easy to understand, clearly visible, and not misleading. A label needs to show energy, fat, saturates, sugars, and salt (see Figure 3.5).

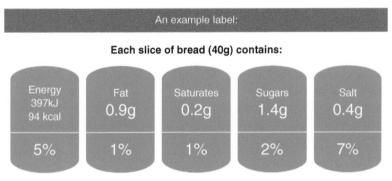

An example label:

Each slice of bread (40g) contains:

| Energy 397kJ 94 kcal | Fat 0.9g | Saturates 0.2g | Sugars 1.4g | Salt 0.4g |
| 5% | 1% | 1% | 2% | 7% |

of an adult's Reference Intake.
Typical values (as sold) per 100g: Energy 993kJ/235kcal

Figure 3.5

ACTIVITY

Food

1 Groups can research into the dietary significance of one of the 5 components of food labels from selected library and Internet resources. Provide a report template so that clear guidance is in place as to the extent, detail and time allotted to the research activity. Limit the discussion to each component's significance in the balanced diet, rather than extending this to include the problems that arise in excess or deficiency.

2 A collection of food labels can provide the numerical basis for analysis and comparison between foods. These may be used to lead to the establishment of food groups. Learners may consider why typical values are expressed 'per 100g'.

3 'Traffic light' labelling is adopted in some supermarkets and can provide a basis for discussion.

What is food?

At its most basic level food can be considered as being the chemical material which is derived from cells and their activities. Food provides chemical energy and nutrients for all living things, and much of environmental science is concerned with feeding relationships throughout the natural world.

What is cooking?

While historians might argue about the antiquity of cooking as a human activity, it is clear that without processing many foods would be unfit for human consumption. Cooking involves combining ingredients and usually applying heat. Broader considerations into the purpose of cooking must acknowledge the aesthetic values of foods and the entertainment value of enjoying a well-prepared meal. The significant creativity that is involved in food preparation is celebrated throughout the media. It is also worth noting the central cultural significance that prepared food has occupied throughout civilisation.

Is pickling a form of cooking?

It might be argued that since heat in most forms of cooking leads to a change in the properties of several of the chemical constituents, while removing pathogens, that pickling might be considered a form of cooking since it achieves the same end.

What is the purpose of cooking?

Food safety is increased by cooking since raw foods (e.g. meat, fish and eggs) harbour food-poisoning bacteria. *Salmonella*, *Listeria* and *Campylobacter* are three of the most common food-poisoning bacteria. Most pathogens are killed at temperatures that are greater than 70°C. Note that refrigeration below 5°C simply slows down bacterial growth. Domestic freezers customarily operate at -18°C, and while bacterial growth is prevented at this temperature, their spores are able to survive and proliferate once food is returned to room temperature, thereby requiring the food to be consumed quickly.

Digestibility is increased by cooking. Many of the chemicals that are contained within foods are not easily digested by the body.

Raw potatoes are fairly unappetising and indeed have poor nutritional qualities since the starch is locked into the potato cells which are surrounded by indigestible cellulose. Cooking breaks the cells up and causes the starch grains to expand and become available to digestive enzymes in the alimentary canal.

ACTIVITY

Popcorn

Cereal crops are grass seeds which are grown for their stored starch. Many of the great crops of the world (e.g. wheat, barley, oats and maize, rice, sorghum, rye) are grass species of global agricultural significance. Maize seeds are surrounded by a hardened, impervious kernel and the result of heat on the starch causes grains to explode.

Make some popcorn. This is more dramatic if it is carried out with popping corn in a pot (remember to use a lid!) rather than microwaved products. Follow the supplier's instructions and take care to avoid opening the pot until all the popping has finished.

Compare the volume of un-popped and popped corn. You might also give some consideration to the dramatic increase in surface area, which at the chemical level increases the opportunities for starch digestion by the enzyme amylase which is released in the mouth and intestines. (Note that nutritionally popcorn is rich in fibre and starch, however the addition of excessive fats and sugars in prepared snacks leads to it becoming a fairly poor food choice.)

Cooking and chemistry

The basis of chemical change can be applied to cooking. Materials change their properties when heat is added or taken away from them. Energy gain changes some materials from solid to liquid, while energy losses change some materials from liquid to solid. These 'changes of state' can be reversible or irreversible. Some of the changes to materials are reversible (e.g. water freezing and melting) and some of the changes to materials are irreversible (e.g. egg white when cooked).

ACTIVITY

State changes

Prepare a classroom display that includes butter, chocolate, sugar, candle, cake mixture, toast, ice cream and jelly. Explore state changes and whether the changes in the materials can be reversed. Butter melts on the table, turns solid in the fridge (Can be changed). Chocolate melts in your mouth, or in your hand, and turns solid on the table (Can be changed). Sugar dissolves in water (Can be reversed by evaporating water). Candle wax melts but turns solid again when the heat is removed (Can be changed, although the flame will turn a small amount of the wax into gases and this is irreversible). Cake mixture turns solid when it is cooked (Heat added and cannot be changed). Toast burns with heat (Cannot be changed). Ice cream melts (Can be changed by returning to freezer). Jelly sets (Quite complicated since preparing jelly involves dissolving and changing the proteins with heat – the main point is that jelly can turn back into a liquid depending on the temperature).

Food preservation

The cells of microorganisms are capable of growing at alarming rates in suitable conditions – they need food, water, and a suitable temperature for growth. Food-poisoning microorganisms release toxins that cause sickness, and we can take several approaches to prevent their growth.

ACTIVITY

Food preservation

Undertake out-of-class learning in a supermarket or the school kitchen to observe food preservation techniques. Look for examples of freezing, drying, pickling, canning and salting.

Since a low temperature slows the microbial growth rate, refrigeration and freezing are useful techniques. Note that neither refrigeration nor freezing will kill microorganisms and so careful food management is required with frozen food and food kept inside a refrigerator.

Salt provides a traditional approach for preventing bacterial growth. Salted fish, beef and ham are common products, and children may be familiar with vacuum packed bacon (salted pork which has been packed to prevent further contamination by microorganisms). (Compare this with prosciutto/iberico ham hanging up in delis, bars and restaurants.)

Dehydration (the removal of water) is effective in food preservation since moisture is required for microbial growth. Dehydrated products include soup and milk.

Cooking kills most microorganisms, but it is not until food has been held at 121°C for at least three minutes that all microorganisms and their spores will be killed. This can be achieved domestically in a pressure cooker and industrially in an autoclave. Canned food is treated in this way and sealed in sterile conditions to prevent contamination. Children should be taught that 'blown' cans must be discarded.

Industrial processes

Biotechnology involves exploiting microorganisms for commercial purposes. Examples exist in food processing, brewing and baking, dairy, agriculture and pharmaceutical industries.

Cells and food preservation

Throughout civilisation humans have learned to use microorganisms to preserve foods. The techniques commonly involve providing conditions that encourage the growth of particular microorganisms which are inoculated into the food in question. Populations of the favourable microorganism compete and prevent the growth of microorganisms that are responsible for food spoilage.

Edibility and palatability

Edibility (i.e. something that is fit to eat) and palatability (i.e. food which is tasty to the individual) are both influenced by cooking. Hunger is a basic human drive that cannot be adequately satiated by inedible materials since these will not provide the body with essential nutrients. Cooking brings about changes in the colour, flavour and texture of foods. Each of these influences the capacity to be digested and how it tastes. When foods are heated significant chemical and physical changes take place. Learners can compare raw and cooked eggs, meat or vegetables.

The cells of microorganisms play an important part in the flavour and aroma formation of many food stuffs. For example, the flavour of mould-ripened cheeses is determined by the breakdown of products of fungi (e.g. *Penicillium roquefortii* in blue cheeses).

Single cell proteins refer to foods that are taken from entire or an extract of microorganisms (e.g. Marmite [from yeast] and Quorn [from mould]). Microorganisms' cells are also used in the commercial production of vitamins, organic acids, colouring, and polysaccharides.

Milk, mutation and civilisation

Milk is a useful nutritional source of protein, lipids, vitamins and minerals, but it has a comparatively short shelf life before bacterial growth leads to its souring. Pasteurisation removes pathogens in large numbers of bacteria but it is not sterile:

> Milk is mostly water but it also contains vitamins, minerals, proteins, and tiny droplets of fat suspended in solution. Fats and proteins are sensitive to changes in the surrounding solution (the milk). (Choudhary, 2012)

Although milk was the principal source of nutrition for humans for their first four or five years until around 10,000 years ago, following weaning, humans lost use of the enzyme that digested lactose (milk sugar), and this resulted in cramps and diarrhoea. A mutation in an individual (probably male, and somewhere in central Europe) left the enzyme switched on, thereby allowing those who inherited this gene to drink milk all their lives. Adults who do not carry the mutation are unable to digest lactose and are described as 'lactose intolerant'. Geneticists have been able to map the spread of this mutation throughout Eurasia. They have also been able to demonstrate other mutations that took place in Africa and the Middle East, but not in the Americas, Australia or the Far East.

The lactase mutation has been closely associated with the way in which civilisation has developed and the domestication of cattle for milk production. In Europe fewer than 5% of the population are lactose intolerant, while in East Asia – where domesticated cattle appeared much later – more than 90% of the population are lactose intolerant.

Milk has a comparatively short shelf life, even when refrigerated. The cells of microorganisms can be used to preserve it for longer periods in solid forms (i.e. as cheese) or as a gel (i.e. yoghurt).

Cheese is a solid food that has been made from the milk of domesticated animals (e.g. cattle, goats or sheep). It is formed by the coagulation of casein protein which additionally traps milk fat. An enormous range and variety of cheeses are available that have different appearances, textures and flavours. Cheese production involves controlled microbiological growth to ensure the quality of the final product.

Curds and whey

When milk sours it separates into liquid and solid components. It has an unpleasant smell due to the breakdown of the nutrients that are contained within it.

Microorganisms can be used to control the souring of milk, competing against potential pathogens, and provide a product which will last for several weeks.

The solid part, the curds, includes proteins, mainly casein, and milk fat. The liquid is mainly water and its main chemical constituent includes dissolved lactose, some protein, and small amounts of minerals.

ACTIVITY

Milk

1   Acids cause milk protein to denature and coagulate. This will lead to separation of the milk solids. Investigate the effect of adding lemon juice to milk. Compare fresh and preserved lemon juice.

2   Rennet is available from chemists. Experiment by adding small volumes (i.e. one or two drops to 50 ml milk) and look at the results after 30 minutes. It is possible to vary the number of drops, the type of milk, and the incubation temperature.

Yoghurt

Yoghurt is produced by the bacterial fermentation of milk. *Fermentation* is a term that is used to describe any process that depends on microorganisms – bacteria and fungi – to convert chemicals into other ones. Its most familiar use is in the fermentation of fruit juices by yeast – a fungus – to produce alcohol.

Milk sugar – lactose – can be metabolised anaerobically by bacteria to produce lactic acid waste. This lactic acid in turn denatures the milk protein – casein – and this causes the milk to set.

Several types of bacteria are used in yoghurt production (e.g. *Lactobacillus, Streptococcus* and *Bifidobacteria*). These are listed on the labels of 'natural' or 'live' yoghurts and can be used as an inoculum/starter to make your own yoghurt).

ACTIVITY

Yoghurt

a) Add two teaspoons of powdered, skimmed milk to 500ml of whole milk. (This increases the protein content of the raw material.)

b) Bring the mixture to a boil over medium heat for 30 seconds, stirring constantly to kill any unwanted bacteria.

c) Cool to 46–60°C.

d) Place one to two teaspoons (5–10ml) of live yoghurt in the cooled milk mixture.

e) Stir this mixture well using a spoon previously sterilised through boiling or left to stand in very hot water.

f) Pour the mixture into sterilised cups or dishes and cover these with aluminium foil.

g) Incubate the mixture at 32–43°C from 9 to 15 hours until a desired firmness is reached.

h) Add jams, sugars, nuts, fruits etc. to add flavour.

Cheese

Pasteurisation is not an essential part of the production process, however food safety legislation requires this in most industrial cheese production. Pasteurisation – invented by the nineteenth-century French scientist Louis Pasteur – was developed to prevent wine and beer souring, and kills most bacteria by heating and cooling the liquid quickly so that the microorganisms are killed and the liquid remains largely unchanged.

Pasteurised milk removes pathogens and has been a significant public health measure.

Starter cultures of particular bacteria are added to the milk, and these convert the lactose sugar into lactic acid which subsequently causes the casein to curdle and create a gel. Rennin (an enzyme) can be added at this stage to accelerate the milk curdling. Rennin was originally taken from calves stomachs but much of it is now obtained from bacteria. The milk is fermented for several hours to create curds and whey.

Cutting the coagulum into cubes assists the separation of the curds from the whey.

Draining and pressing will continue to separate the curds from the whey.

Salting adds flavour and preserving qualities to the cheese.

Ripening and storing cheese will allow it to mature if required and develop flavour.

Additional microorganisms can be added at various stages (e.g. *Penicillium* mould is spiked into blue cheeses, and propionic acid bacteria are added to make the bubbles in Swiss cheese).

ACTIVITY

Cottage cheese

Choose a recipe from the Internet to make cottage cheese. It is simple to make and mainly involves the separation of curds and whey. The curds have a mild flavour and remain moist with whey.

Growth, development and repair

New cells are made by cell division. A single cell divides into two smaller ones. These each mature and return to their full size and subsequently divide into four cells. This process of division and enlargement will continue and make eight cells, sixteen cells and so on. *Multicellular* organisms grow by making more cells which subsequently get bigger.

ACTIVITY

Seeds

1 Collect and dry seeds from a variety of fruits – the more unusual the better! Can the class grow oranges, kidney beans, grass, horse chestnuts (conkers), avocados, and kiwi fruits?

2 Investigate 'What might affect' germination in selected seeds (e.g. cress seeds on damp kitchen paper).

Cell division enables growth of the organism, its development, maintenance and repair. In humans the single cell that was produced at fertilisation multiplies into trillions (10^{12}) of cells by the teenage years.

While the numbers of cells increase in multicellular organisms they also specialise to perform different functions. Although skin, liver, muscle and nerve cells have different appearances and carry out different purposes in the body, they all contain identical genetic material.

Cell division takes place throughout our lives to repair damaged parts and replace damaged and worn-out cells. When you cut your skin, rapid cell division on either side of the wound replaces the damaged tissue and quickly restores it to full health. It is perhaps surprising to learn that broken bones are repaired by division of the specialised bone cells. Our skin is constantly regenerating and rubbing off at the surface. Severe damage to our bodies cannot be repaired however, for example the loss of a limb. Many animals though are able to regenerate lost limbs (e.g. salamanders, crabs and starfish).

Tissue culture

Tissue culture is an important scientific and commercial process that relies on cell division and the production of identical copies of cells, known as *clones*. Tissue culture involves growing plant and animal cells and tissues in sterile laboratories that provide carefully controlled conditions (such as light, temperature, pH, hormones and nutrients) to ensure that growth takes place.

Many plant cells and tissues are able to grow into a whole new plant. This *totipotency* can sometimes be a nuisance as tiny pieces of garden weeds, such as dandelions and couch grass, are able to grow back into entire plants. These plants are successful weeds since they are capable of growing new shoots and roots from the tiniest fragments that are left in the soil that can grow into whole new plants/weeds.

Photo 3.2 *A tiny fragment of weed left in the ground can grow roots and regenerate a whole new plant*

© N Souter

ACTIVITY

Growth

1   Carrot tops (and several other root crop tops) can be grown on a window sill. Cells grow and differentiate into new shoots and roots. Ensure the tops are always standing in water but are not flooded by it, and that the growing green tissues are standing in the air.

2   Collect the fragments of weed roots (e.g. couch grass or dandelions), cut them into 2cm sections, and plant the segments in damp potting compost or vermiculite (available from garden centres). Look for regrowth during the next few weeks.

Tissue culture is a very quick method of providing clones of plants which have useful traits – for example, attractive flowers, tasty fruits, high productivity, or resistance to disease. It also allows scientists to manipulate the genes that are present in individual cells in a process called *genetic modification*. Cloning is also important in the conservation of rare plants which may have been removed from their natural habitats.

Losing control

Cell division is controlled by living things so that growth and repair take place in an organised way. Sometimes, however, the process does go wrong and a *tumour* can result.

Photo 3.3   *A witch's broom gall on birch tree*

© N Souter

The unusual growth in the photo is a tumour which has been caused by a fungus. The extra growth is brought about by uncontrolled cell division but it does not harm the tree. Warts and verrucas are examples of tumours that are caused when our skin is infected with the human papilloma virus (HPV). These skin tumours can be removed quite easily. Different kinds of HPV are associated with cervical cancer. A *cancer* causes disease that spreads to different parts of the body. Today girls aged 12 or 13 are offered HPV vaccination to protect them from cervical cancer which can be fatal.

Scientists are unsure how many cancers are brought about but they do understand that changes in DNA lead to changes in cell growth and the formation of tumours that can spread in the body and become cancerous. Several factors – including chemicals, radiation, viruses, inheritance and age – can influence cancer development.

Stem cells

Stem cells are unspecialised 'master cells'. They are the cells from which all others mature. Their role is to make new cells. Without stem cell development, growth and repair could not take place.

Stem cells are something like a production line since they manufacture more stem cells or develop into specialised ones (e.g. nerve cells, skin cells, kidney cells, etc.). Specialised cells *differentiate* by having different appearances and carrying out particular functions inside tissues.

Stem cells are found in embryos, the foetus and the umbilical cord, as well as in adults. Stem cells in the bone marrow differentiate to form red and white blood cells. Your skin has a range of stem cells that ensure it is constantly replaced and kept in good condition.

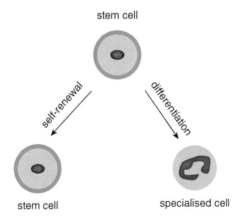

Figure 3.6

Living factories

Cells can be described as chemical factories. All living processes are controlled by chemical reactions which take place in cells. These chemical reactions occur very quickly because cells contain *enzymes*. Cells and the enzymes found within them can be used to manufacture useful materials. The cells of bacteria and fungi are especially useful.

Yeast industries

Fermentation forms the basis of both the brewing industry and the baking industry. Most brewing is carried out using varieties of the yeast species *Saccharomyces cerevisiae*. Yeast grows naturally on the surface of fruits and helps the process of decay. Alcohol can be produced from almost any sugar source. Beer is made from maltose sugar – the product of starch digestion in germinating barley. Beer is often flavoured with hops. Wines are made from fruits, usually grapes. Spirits like whisky and gin are made by preparing a simple fermentation and then distilling the alcohol to concentrate it.

Baking bread

The carbon dioxide formed when yeast reacts with sugar is used in the baking industry to make bread rise. Bread is one of the most common *staple* foodstuffs in the world. Bread dough includes the following ingredients: flour, water, yeast, salt, oil, and sugar.

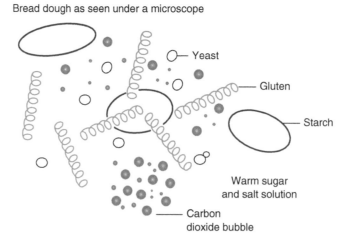

Bread dough as seen under a microscope

Yeast

Gluten

Starch

Warm sugar
and salt solution

Carbon
dioxide bubble

Figure 3.7

The flour provides starch and gluten – a protein that provides the elastic properties of the dough as it rises and gives bread its 'spongy' texture. Sugar is added to supply a source of energy for the yeast which is unable to digest starch. Salt is added both for flavouring and to provide the correct environment for the yeast while it grows and respires.

ACTIVITY

Dough

1 Ask the children to describe the properties of a golf ball-sized piece of bread dough while they knead it.
2 Investigate bread dough rising. Spray oil inside a glass tumbler and press a golf ball-sized piece of bread dough into the bottom. Mark the level of the dough with an elastic band outside the glass. Cover the glass with cling film or damp paper towelling to prevent it drying out and forming a crust. Incubate in a warm place. Dough rising can be recorded photographically. It is possible to vary the following factors: water, incubation temperature, salt, sugar, flour, and yeast. Observe the effect each change has on how much the dough rises and how fast the process happens.

Ginger beer

Ginger beer has been a popular drink for more than two hundred years. Alcoholic and non-alcoholic versions are available in the following recipe that has been taken from Microbiology Online (www.microbiologyonline.org.uk/). Teachers may prefer to place an emphasis on production of the starter 'plant' only, thereby avoiding unethical practices associated with alcohol ingestion.

Libby Riley's recipe for ginger beer

(This recipe makes five litres or nine pints)

Ingredients

- 25g (1oz) fresh yeast or 15g (1/2oz) of dried yeast
- 1kg (21/4lb) sugar
- 40ml (8tsp) ground ginger
- Juice of two lemons
- Water

Method

Starter 'plant'

Put the yeast into a large clean jar. Pour in 275ml (10fl.oz) warm water. Stir in 10ml (2tsp) sugar and 10ml (2tsp) ground ginger. Cover and leave in a warm place for 24 hours. This is the starter 'plant'.

Feeding the 'plant'

1   On each of the following six days feed the 'plant' with 5ml (1tsp) sugar and 5ml (1tsp) ground ginger. Stir and cover the jar each time.
2   After the last addition leave the solution to stand covered for another 24 hours.
3   Line a sieve with muslin. Strain the solution, reserving both the liquid and the sediment.
4   Over a low heat dissolve 900g (2lb) sugar in 575ml (1pt) water. Stir well.
5   When the sugar has dissolved bring to the boil and boil for three minutes.
6   Pour the syrup into a large bowl and stir in the lemon juice and liquid from the plant.
7   Dilute with 3.5l (6pt) water, stir well and pour into clean, rinsed out bottles. Secure the bottles with corks (do not use screwtops as yeast produces carbon dioxide as it ferments liquids which may cause the bottles to explode).
8   Store for at least one week before using.

Use half of the reserved sediment as the new 'plant' to start the next batch of ginger beer. Give the other half to a friend so they can make their own.

Time

5–10 minutes for eight days, then one week's storage.

Reprinted with permission from the Microbiology Society

Bacterial industries

The metabolism of bacteria can be used to provide useful materials, mainly in the food and pharmacy industries. The controlled souring of milk by bacteria lies at the basis of yoghurt and cheese production. Other products include the following.

Silage

Winter food for cattle often includes silage. Silage is prepared by harvesting grass several times during the growing season and packing it beneath waterproof materials, such as thick plastic sheeting. The silage remains wet, and in the absence of oxygen bacteria release acids to 'pickle' the grass and thus preserve it.

Sauerkraut and kimchi

Sauerkraut and kimchi are popular in Europe and Korea. Each of these delicious foods is prepared by the fermentation of the plant material by bacteria to release acids that preserve the food and prevent bacterial contamination, moulds etc.

Antibiotics

Blue mould in cheese is a fungus. Species of this mould gave the world its first antibiotic. Although there are now over 100 different types of antibiotics, development of new types of antibiotic continues all the time since bacteria mutate and become resistant to existing ones.

ACTIVITY

Fungi use

Invite group researching of 'Using fungi' in the upper school (e.g. yeast, penicillin and food, penicillin and antibiotics, Quorn). Learners should make a slide presentation. This should be structured to provide five slides with suitable illustrations and information.

Biological washing powders

Washing powders contain detergents to help clean our clothes. Biological washing powders contain added enzymes that are extracted from bacteria. The enzymes in the washing powder help to break down some of the chemicals that may stain our clothes, such as egg yolk or sweat. These enzymes are extracted from bacterial cells. Amylase enzymes remove starch-based materials. Protease enzymes can help to remove protein stains such as gravy or egg.

ACTIVITY

Stains

Compare the effects of soaking stained fabric samples in biological and non-biological detergents.

Making use of enzymes

Enzymes have been used in cheese production for hundreds of years. They are widely used in industry wherever precise chemical changes are required (e.g. in food processing to release sugars from starch, tenderise meat, and clarify fruit juices). Contact lens cleaners contain protein-digesting enzymes to help prevent infections. Enzymes are even used to produce soft-centred chocolates. Biofuel and detergent production also involves enzymes.

Enzymes are involved in DNA technology and are essential research tools, especially in genetic engineering and pharmacology (medicine production).

Medical uses of enzymes

Enzymes are used in medicine to treat blood clots and some types of tumour. The development of enzyme-based treatments can be difficult and expensive as enzymes can be hard to extract in a pure form from cells. Enzymes are also used in some diagnostic tests, such as testing for sugar in urine to help diagnose diabetes.

Gene therapy

Inherited diseases frequently involve faulty genes. Gene therapy involves the use of viruses to carry corrective genes into patients. People with the genetic disease cystic fibrosis (CF) lack the correct genetic information and their lungs become clogged by sticky mucus. Treatment involves daily physiotherapy to dislodge the mucus. It is a serious inherited disease which shortens the lives of sufferers. Experimental approaches in gene therapy have been used to treat cystic fibrosis. Scientists have attempted to deliver DNA into the cells of a CF patient using a virus as a vector (carrier). Viruses were used for this purpose because they are able to invade cells. Normally when they do this they cause disease, so scientists developed virus vectors that would carry information but not cause disease. The safety of introducing viruses to the lungs of individuals who were already very ill was a cause for concern among the scientific team. What if the virus vector caused disease? What if the individual's immune system attacked the virus?

Gene therapy is still causing debate but new therapies are being developed all the time for diseases such as Parkinson's. Once the difficulties regarding safety have been overcome it may, in the future, be a very useful treatment for many diseases.

Environmental biotechnology

Waste treatment

Human activity and urbanisation have led to the accumulation of organic waste which requires removal. Decomposers throughout all food webs break down organic materials from dead living things or their waste materials, and return the mineral content to the environment. Waste treatment exploits this recycling stage.

ACTIVITY

Decomposition

Leave an orange or another piece of fruit to decompose. Make a photographic record of the changes that take place over several weeks.

Photo 3.4

© N Souter

ACTIVITY

'Fishing for microbes'

It's easy to go 'fishing for microbes' almost anywhere, and you can take your class on a 'fishing trip' by getting them to leave plates of bread, cheese or fruit exposed to the air, in a safe place, and wait for a few days or weeks.

What the pupils will see is the result of the microbes in the air settling on the food and feeding off the sugars and proteins in it, eventually developing fruiting bodies that we call mould.

You might carry out an investigation to observe bread decomposing over several weeks. Start off by cutting a disc from a fresh slice of bread using a pastry cutter.

(Continued)

(Continued)

Leave the disc of bread exposed for 10 minutes and then place it in a resealable bag. Write the date on the bag. Do not open the bag again but keep it for comparison for the next few weeks.

Each week you should prepare a new disc in exactly the same way as before. Keep previous ones to build up a sequence as a class display. After a number of weeks these samples can be used to add to a display of decomposing bread. You might also take photographs to record the changes that take place over several weeks.

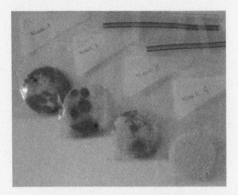

Photo 3.5
© N Souter

Composting

Legislation, designed to reduce the extent of organic waste that enters landfill sites, requires collection from towns and villages. However collection is only the start of the recycling process.

A domestic compost heap is similar in organisation to the model shown below (i.e. layers of garden waste including weeds, grass cuttings and dead plant material are built up and inoculated with layers of topsoil that include bacteria, fungi, and a host of tiny animals). The respiration produces heat that kills the weeds and accelerates the process of decay. Food chains become established so that the material is broken down into humus that can be returned to the soil.

The garden waste that is collected from homes must be treated in the same way but on an industrial scale. The first stage involves sieving the materials to remove items which cannot be composted. The waste is shredded and piled in rows where decomposition takes place. Frequent turning every five to ten days introduces air which helps the breakdown. After about 20 weeks the compost will be free of weeds and rich in organic material and plant nutrients that can be returned to gardens, horticulture or farming.

ACTIVITY

Compost

You might take the opportunity to set up a compost column in your classroom. This will allow learners to see the different stages in the decay of organic waste. Access the science and plants for schools (SAPS) website for full details on how to make the column (www.saps.org.uk).

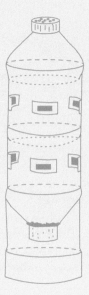

Figure 3.8

Waste water treatment

Sewage treatment involves microorganisms that break down organic waste and use it for their own growth. Food webs are established which purify the water and remove pathogens. The same process takes place whether it is inside a septic tank inside a rural property or in a massive urban sewage works.

ACTIVITY

Filtering

Prepare a model filter bed using a drinks bottle, with the bottom removed, to show the way in which particles can be removed (by filtering) from water as part of the purification process.

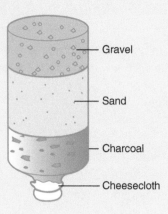

Figure 3.9

Food waste

Food waste is additionally hazardous since it can attract vermin which can in turn spread disease. Industrial treatment involves food being digested by microorganisms inside sealed containers in the absence of oxygen (anaerobic digestion). The food is turned into methane gas and humus. The biogas produced in this way can be used to heat the container, generate electricity, or power vehicles. When all the biogas has been produced, the remaining material, humus, can be used as fertiliser because it still contains useful nutrients.

Energy capture

Renewable energy sources produce electricity or heat without depleting our planet's limited resources. *Biorenewables*, also known as *biomass*, are produced by living things. They include agricultural crops and wood produced by photosynthesis, as well as animal waste.

Biorenewables may generate heat from the direct use of the flammable materials as a fuel or they may be converted into liquids and gases. Ethanol, a biofuel, is manufactured by converting the cellulose or starch in the biomass into sugar which

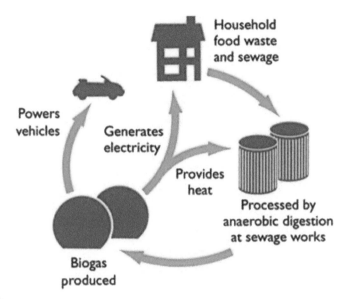

Figure 3.10

Reproduced with permission from Geneco

is then fermented into alcohol (ethanol). Ethanol can be used to power cars. It is produced by fermenting the sugar or starch in sugar cane, sugar beet, wheat, maize or wood pulp into ethanol.

Biodiesel is produced by mixing alcohol with oils that have been extracted from plants such as rapeseed, soybean, or from waste vegetable oils. Crops grown as fuels include food crops such as sugar cane and corn, grasses, and rapidly growing trees.

Conflict may arise in the future between food and fuel production on finite land resources.

Biogas is produced from decomposing animal or plant waste. It has a high methane content.

Anaerobic digestion is carried out by bacteria. It is used to break down sewage, food waste and other organic materials to produce methane gas. The solid remains can be spread on fields as fertiliser.

ACTIVITY

Energy recapture

1 Visit a domestic recycling centre
2 Invite a speaker from the local authority.

Transgenic plants

The cells of transgenic organisms contain a gene or genes that have been been genetically engineered, to confer new characteristics. Transgenic varieties are referred to as genetically modified organisms (GMO), although in fact all domestic animals and crops have been genetically modified by selective breeding from their wild forms over extended time periods. (See Chapter 6.)

Summary

That all living things are made from cells is one of the most important ideas in life science. Their discovery a little over three hundred and fifty years ago, by Robert Hooke around the time of the great Fire of London by using simple microscopes, has led to today's breathtaking insights about their complexity in chemical and physiological responsiveness. Modern resources, especially the advent of low-cost electronic microscopes, have revolutionised our capacity to look more closely at small things and we encourage their use with children. Cell biochemistry and the 'big' processes of respiration and photosynthesis underpin the survival of all living things on this planet. This chapter has introduced the study of cellular processes including their growth, reproduction, and survival. We have examined the nature of food as well as the usefulness to which cells, especially microorganisms, have been put in food, agriculture and pharmaceutical industries.

REFLECTION POINTS

- Make a mind map to show the relationships that exist between the various chemicals that occur in nature.
- How many types of cooking can you name?
  - Compare and contrast two of these approaches.
  - (Your answers might include reference to the cooking matrix, temperatures, suitable foods, product texture and appearance.)
- Which additional public health measures take place to ensure milk is free of major pathogens (e.g. bacteria that cause tuberculosis and brucellosis).
- Compare and contrast 'tumour' and 'cancer'.

4.
PLANTS

Learning Objectives

Having read this chapter you should be able to:

- extend your insight into the biodiversity of plants
- identify plant parts
- illustrate how plants have benefited society
- grow plants from seeds and cuttings
- explore sexual reproduction in flowers
- investigate food and oxygen production
- find out what a plant 'knows' in order to respond to the environment
- look at adaptations in plants to meet environmental variation

Overview - plant biodiversity

What are plants?

Classification is a notably complex area of life science. As with most aspects related to the classification of living things, scientific nuances necessitate providing a working definition. Although exceptions exist, a working definition is that plants are terrestrial or aquatic, photosynthetic organisms whose cells are surrounded by cellulose walls.

Plants provide nourishment. Plants, and photosynthetic protists, provide atmospheric oxygen. Alongside fungi, plants provide the food that sustains us either directly or indirectly. Prominent plants frequently define ecological communities. 'Oak forest' or 'grassland' provides some insight into the entire ecology, including physical and biotic features.

Plants are classified into 12 divisions, and these are based largely on reproductive characteristics and into those that involve spores, naked seeds, or covered seeds; tissue structure into non-vascular (e.g. mosses) and vascular plants (all others); or by growth forms (e.g. into mosses, ferns, shrubs and vines, trees, and herbs). Further details on non-flowering plants are provided in Chapter 1.

Plant parts

Plants live mainly on the land and only have five simple components – roots, stems, leaves, buds and flowers (and the last two are really modified leaves).

The variety of adaptations and appearances of land plants are however not at all simple since they are capable of creating an extraordinary diversity that exploits every ecological niche. Angiosperms are adapted to drought and flood, light and shade, acid and alkaline conditions, high winds and rapid currents, mineral shortage

Above ground level plants absorb Carbon Dioxide from the atmosphere.

Leaves trap solar energy for photosynthesis.

_____

Below ground level plants absorb water and dissolved minerals.

Figure 4.1

or high levels of salt, and every possible condition on the planet. Their sizes, shapes and appearances are worth further study. The range of sizes of flowering plants alone is staggering – compare duckweed with a mature oak tree. The masses of these plants range from a few milligrams to 400,000 kilograms!

ACTIVITY

Collage

1. Collect and press leaves and/or wildflowers from inside the school grounds to build a resource library for comparison with future classes. Prepare a collage of pressed autumn leaves.
2. Invite groups to research one of the following:

   The largest land plant
   The smallest land plant
   The largest flower
   The fastest-growing plant

Non-flowering plants

Mosses and allies (*Bryophytes*)

Mosses, liverworts, and hornworts reproduce sexually via the production of spores. They can be found growing on the ground, on rocks, and on other plants. They are non-vascular (i.e. they have no internal transport system). Since they cannot transport fluids through their bodies they must rely on surrounding moisture. Although they are individually small the mosses are very important members of our ecosystem. They are significant pioneers that help colonisation, assist humus production, and prevent erosion.

Although fascinating biodiversity exists within the bryophytes one group in particular, the *Sphagnum* species, are of wider interest due to their ecology, history, energy production, horticultural use, and topical conservation interest.

Peat bogs are formed by an accumulation of *Sphagnum* moss, a water absorbent moss that fossilises to form peat. Peat bogs are acidic and bacteria cannot grow there. This prevents decay.

Since it is sterile and absorbent, the remains of pollen that is trapped in each layer of peat can reveal information about the climate and the variety of plants at any time in history since the formation of the peat bog. This usually refers to the period since the last ice age. Occasionally ancient bodies have been found preserved in the peat. 'Pete Marsh' was the trivial name given to one such body in Cheshire, England, after it was found in 1984. The 'Tollund' man in Denmark also allowed forensic analysis of the ancient diet (c. 350 years BCE).

Sphagnum's sterile and absorbent qualities have allowed it to be utilised in the past by humans as toilet paper, and as wound dressings from ancient times until as recently as the First World War.

Sphagnum can be used to line flowering baskets and hold water. It can also be used to produce potting compost and as a soil conditioner. Peat is a fuel that is used sustainably in rural communities, however it also used to power electricity-generating stations in the Irish republic, Scandinavia, and Russia. Since *Sphagnum* is a finite resource such industrial uses can threaten both the plant and its habitat.

ACTIVITY

Sphagnum moss

1   Explore the water-holding capacity of *Sphagnum* moss. (This activity should only be carried out if small quantities, say no more than a handful, can be collected locally.)
2   Research the risk of habitat destruction due to *Sphagnum* removal. (Horticultural uses and/or energy production.)

Ferns and allies (*Pteridophytes*)

Ferns, horsetails and club mosses reproduce from spores rather than seeds. They have a vascular system to transport fluids. In some areas young ferns are harvested for use as a vegetable (i.e. fiddlehead greens). Horsetails' abrasive qualities lent to it being used for scouring pots in ancient times. Fossilised horsetails are frequently found in coal reserves, showing that it grew 350 million years ago.

Horsetail (*Equisetum arvense*) has been used as a herbal remedy since Greek and Roman times to stop bleeding, heal ulcers, and treat tuberculosis. It has also been used as a diuretic and may provide relief from kidney disease. Its full medicinal potential has not been extensively studied.

ACTIVITY

Ferns/horsetails

1 Press fern leaves (fronds) and examine the lower surface for spore production.
2 Collect some horsetails. Direct experience of these non-flowering plants will permit the children to feel their unusual surface, which includes silicon materials, and the resultant rough texture. Let the children pull these apart and see how they piece together, almost like Lego.
3 Make a giant horsetail from sugar paper cylinders using cylinders of progressively larger diameter recycled paper. Serrated edges will emphasise the 'Lego' effect.

Conifers and allies (*Gymnosperms*)

Gymnosperms (i.e. pines, firs, spruces, cedars, junipers, and yew) reproduce from seeds (i.e. an embryonic plant with a protective covering) instead of spores. The seeds are 'naked' in *Gymnosperms*. These are compared to higher plants where the seeds are surrounded by an ovary. *Gymnosperm* seeds are usually produced inside a cone, hence the name 'conifer'. However a few (e.g. yew seeds) are produced in something that resembles a berry.

Conifer identification is helped by the presence of cones. Male and female cones are involved in sexual reproduction and following fertilisation the female cones mature to disperse the seeds. *Gymnosperm* leaves are frequently needle-like and this reduces water loss, helping conifers to colonise habitats with water availability problems e.g. coastal sand belt and frozen areas.

They are exploited principally for timber. Other uses include the extraction of essential oil that is used for a variety of purposes, including aromatherapy; pine nuts, used in pesto; and juniper berries, used for flavouring gin. Yew seeds are extremely poisonous.

Flowering plants (*Angiosperms*)

Two divisions

Angiosperms, the flowering plants, represent the most highly evolved members of the plant kingdom. Angiosperm seeds grow inside the ovary which lies at the heart of the flower, its reproductive structure. Following fertilisation the ovary swells to become a fruit. There are two groups of *Angiosperms* – dicots and monocots. These names are derived from the construction of the seed which may have one or two 'seed leaves' (i.e. the cotyledon(s)). Several differences exist that assist *Angiosperm* classification (see Table 4.1).

Table 4.1

	Dicot	Monocot
Seed leaves	2	1
Leaf veins	Branched	Parallel
Flowers	4 or 5 petals (or multiples)	3 petals (or multiples)
Vascular system	Arranged in rings	Scattered through stem
Secondary growth	Frequently present	Absent
Tissues	Herbaceous and woody	Herbaceous
Examples	Buttercups, daisies, roses, carrots and peas	Grass, cereals, lilies, tulips, bananas

Dicots

The vast majority of plants, around 200,000 species, are dicots. Most trees, shrubs, vines, and flowers belong to this group. Many economically significant plants are dicots (e.g. those that are grown to produce foods, fabrics, medicine, housing, timber, and used as an energy source).

Monocots

Although only around 15% of all Angiosperms are monocots they include many examples that are economically significant (e.g. food from cereals, animal fodder from grass in meadows). Many garden plants are monocots (e.g. lilies, irises, orchids).

Their growth always starts with one seed leaf and then these grow into mature, photosynthetic leaves which are supplied with water and nutrients from the soil through the characteristically unbranched, parallel veins.

ACTIVITY

Leaf veins

1 Compare spring onions with spinach leaves. Look for the parallel veins and the branching or net of veins. (This is a significant feature, used in classifying flowering plants.)

2 Build a model water lily plant with inexpensive flat sponges cut into leaf shapes, with some branching veins (dicot) drawn with a permanent marker from the point of attachment of each stalk.

Figure 4.2

Float the 'leaves' in a basin or fish tank. String the leaves to a weighted anchor to model the water lily stems. These will model the buoyant nature of water lily leaves that include air pockets to help the pads float on the water surface.

3 Visit a park or botanical gardens and look for examples of monocots and dicots and for biodiversity within these groups.

Have you used a plant today?

All life on Earth depends on the energy flow within food webs; atmospheric oxygen has been produced by the remarkable process of photosynthesis. However plants' significance to humans extends well beyond their position in food webs, or their need for oxygen uptake, and is bound up with the development of our civilisation. Technology has found uses for all plant parts.

Food

Grass *seeds* provide many of the world's staple foods (e.g. cereals and rice). They are frequently milled to create flour. The seeds from the pea family (*Fabaceae*) meet significant nutritional needs in several parts of the world, while seeds from other plant species provide enrichment to our quality of life (e.g. coffee, chocolate and many spices). Fresh vegetable *leaves* (e.g. cabbage and spinach) provide essential dietary fibre and nutrients. Leaves from basil and coriander for example add significantly to the flavour of food. Dried leaves also have a variety of uses and are frequently infused in boiling water to provide a hot drink (e.g. tea). The *roots* of several biennial plants have evolved to store food to support growth in the second season (e.g. carrots, turnips and beetroot). Potatoes perform the same function although they are modified underground stems (tubers). *Flowers* have their uses as food as well (e.g. zucchini in tempura batter, or various flowers in salads such as nasturtiums or violas). Saffron (crocus stamens) is one of the world's most expensive food materials. All sorts of *fruits* are also eaten whole, can be cooked, and can be used to add flavouring. Fruits botanically refer to the part of the plant that is derived from the ovary. The purpose of the fruit is to provide conditions for seed formation and their subsequent dispersal. The answer to the old problem 'is a tomato fruit or vegetable?' should therefore be straightforward to answer in botanical terms. Think of the variety of 'fruits' that exist, ranging from soft fruits to apples and oranges and on to nuts. Plant *sap* can be used to provide fruit juices and may also be concentrated (e.g. maple syrup) or crystallised (e.g. cane or beet sugar). Oils can be extracted from most plant parts but mainly from their *seeds*. Oils can be used for cooking and heating. These can be extracted by applying pressure, or by using solvents, or by distillation. Olive oil is traditionally extracted in presses. The efficiency of extracting oil from rape seed for example is significantly increased with the use of organic solvents. Essential oils are extracted from flowers (e.g. lavender, roses), leaves (e.g. mint and eucalyptus) or flower buds (e.g. cloves) by distillation. They are then used as food flavouring, in cosmetics, and in the pharmaceutical industry.

Clothing

The fibres from cotton seeds, and from the stems of flax and hemp plants, can be spun into threads and subsequently woven into cotton, linen and hemp fabric. Rubber has been widely used for waterproofing as well as for providing soles for shoes. Rubber tree sap is extracted and treated chemically to provide this familiar and versatile material.

Medicine

Plants have been used throughout history as a source of medicine, and even today more than three quarters of the world's population depends on remedies from locally collected plants. Traditional Chinese medicine uses around 5000 different species. Plants' importance cannot be underestimated since more than 40% of prescription medication comes from plant extracts or synthetic copies.

Aspirin for example is one of the world's most successful, a widely used medicine which reduces pain, fever and inflammation, and is also used to prevent heart attacks. Aspirin is derived from salicin which is extracted from willow (*Salix* spp.) bark. Salicylic acid has been synthesised by the German drug giant Bayer since 1897, the same year that Marconi patented the radio.

Around 10,000 tonnes of Cinchona bark is harvested each year to extract quinine and quinidine which are both effective against the malaria parasite. Scientists continue to search for new medications from plant materials, as well as looking at the properties of ancient ones (e.g. artemisinin is a successful antimalarial drug that has been extracted from *Artemisia annua* in traditional Chinese medicine for the last 1000 years).

Housing

Timber is widely used in all aspects of construction, providing a frame, beams, and flooring for example. Timber can also be utilised in a variety of ways for decorative and insulating purposes. Jute fibres are spun into hessian, the backing material for carpeting.

Entertainment

Wood pulp is processed into paper which is turned into newspapers, books and magazines. Many musical instruments (e.g. stringed, percussion, wind) are constructed from timber, gourds, and a range of other plant materials.

Consumer goods

The perfume industry relies on a wide range of plant extracts, and essential oils that provide attractive odours and scents, in the production of perfumes, shampoos, deodorants and soaps. The engineering, decorative and textural properties of wood also make it a preferred material for furniture and home fittings.

Travel

The development of the pneumatic tyre, initially for bicycles – and subsequently for larger vehicles – has been a significant use for rubber. Fermentation products and other processed plant material are being used in the development and extension of biofuels, which can be used for heat and light as well as for providing the energy for vehicles.

Recreation and sport

Turf is the matted surface layer of grass, soil and roots which provides suitable conditions for many sports (e.g. soccer, field hockey, rugby and golf), since it is durable, absorbs impact, and is comparatively easy to maintain. Wood has been extensively used in sport (e.g. for bats, hockey sticks, snooker cues, goalposts). Rubber is also extensively used in sport in the production of footwear, waterproof material, and balls in particular.

Plant growth

We have looked at aspects related to growth and development in Chapter 3 on cell biology, and established that growth is brought about by a combination of increasing cell numbers by division and their enlargement through water uptake and the production of new cellular chemicals. Growth in plants takes place in favourable conditions and increases the mass of chemicals through the uptake of carbon dioxide during photosynthesis. Growth in plants differs significantly from growth in animals since it originates from specialised groups of cells called meristems which are capable of cell division. Meristems appear at the tips of roots and shoots to increase the length of these organs, and around the perimeter of dicots which leads to an increase in girth.

Typical plant life cycle

1. *Seed*
   Seeds are able to survive for extended periods in suitable conditions. They germinate in the presence of moisture, air and warmth.
2. *Germinated seed*
   The energy for early growth is brought about by the breakdown of food reserves in the seed. Shoots grow upwards and roots grow downwards since they are able to detect gravity.
3. *Seeding with first shoot and first root*
   Leaves emerge from the shoot and as soon as they reach the light make chlorophyll (turn green), start making food, and are no longer dependent on the food reserves contained within the seed.
4. *Mature plant*
   Once the plant matures it is able to produce flowers which attract insects which in turn transfer pollen. Fertilisation follows and this leads to the development of fruit.
5. *Broad bean seeds*
   As with all the other plants in the *Fabaceae* these develop inside pods (e.g. beans, peas, sweet peas and gorse). The seeds are typically dispersed ballistically (i.e. the outer coat of the pod dries and ultimately splits, firing the seeds so they land at a distance from the parent plant).

Broad bean life cycle

Although it might seem like a 'chicken and egg' argument we believe that it makes sense to start with growth rather than sexual reproduction.

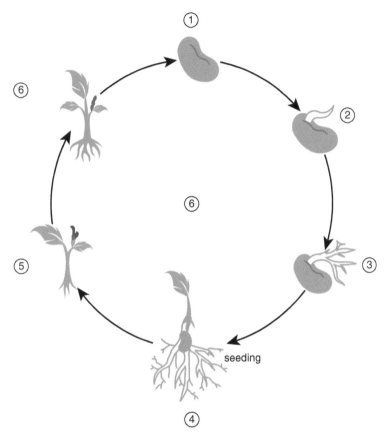

Figure 4.3

Growing plants from their seeds

Seeds are comparatively obvious structures and children can test their ideas by trying to grow them. Initial biochemical activity triggers rapid cell division in the embryo that in turn leads to early growth and the emergence of the first shoots and roots. Ongoing growth and development results in maturation, flowering and the continuation of the plant life cycle.

Food containers, lined with three or four layers of kitchen towel, provide excellent conditions for experiments on germination. Capillary matting, commonly used in greenhouses, also provides a useful reservoir for water. The paper should be soaked but any surplus water should be poured off. The container's lid prevents the evaporation of water and can readily be removed at daily intervals to observe seeds' growth progress (e.g. mustard or cress).

Seed structure

Since seeds are the product of sexual reproduction in flowering plants and contain genetic information that has come from two parent plants, it follows that each seed is genetically unique. Development of the seed is controlled by the genetic information that is found in the nucleus of the cells in the embryo.

Although seeds vary enormously in size and appearance, they do have common features: they include an *embryo* which contains tissues that will develop into shoots and roots; they include a *food store* that provides chemical energy from the early growth of the plant; and they are covered by a *seed coat* – a layer that can protect them from their environment (e.g. decomposers, predators, freezing and soil abrasion).

Photo 4.1
© N Souter

The world's largest seed comes from the Coco de Mer or double coconut, and can weigh as much as 20kg. The smallest seeds are produced by orchids. Their seeds are so small that they can only be seen with a microscope. One seed can weigh as little as 0.8µg (micrograms).

ACTIVITY

Broad bean seeds

Soak enough broad bean seeds (butter beans provide a poor alternative, kidney beans will show the same features but are on the small side) overnight so that there are plenty for all. Dissect the seeds with the class. (No need for sharp knives here as they should slide apart.) Note the leathery waterproof outer layer and the two seed halves, and carefully examine the embryo with hand lenses. Make sure each learner is able to use the lens properly by holding it close to their eye and bringing the object into focus.

Fruit formation and seed dispersal

In flowering plants (*Angiosperms*) fruits form around seeds to assist their dispersal. Dispersal is necessary to ensure that the species colonises new areas and also for its survival. Seed dispersal uses a number of different strategies: via animals (internally or externally); the wind; occasionally by water; or even by the plants themselves.

Animals make a major contribution because when they eat the whole fruit, including the seeds – the fleshy parts of the fruit are digested within the animal's gut but the seeds are not. The seeds pass out relatively unharmed with the animal's waste. Seeds then have the bonus of being deposited alongside moisture and fertiliser. Animals may also act as dispersal agents when seeds become attached to their coats. Pupils will doubtless be familiar with 'sticky willies' (i.e. cleavers or goosegrass).

Other methods of dispersal involve wind and water. Light seeds (e.g. dandelion) have a 'parachute' of hairs to encourage dispersal – sycamore 'helicopters' can provide entertainment in the autumn. Seeds also disperse by being released into water (e.g. coconuts from tropical islands, or the problematic Himalayan Balsam).

Self-dispersal happens when the pod which encloses the seeds dries out and splits to propel the seeds at a distance with a small explosion (e.g. pea plants or gorse pods).

ACTIVITY

Seed dispersal

1   Learners collect seeds from their own garden/area/school grounds and display these in class or bring information together to survey dispersal techniques.
2   Provide a selection of seeds to show the variety of dispersal techniques (wind, animal and water). Prepare display cards with details of the name, type and mechanism of each of these. 'Unknown' specimens should be included in the display – supply the names and ask the children to speculate about the remaining information (as above).

Apples and seeds

Was an apple a flower at one time? The relationship between flowers and fruits is not an obvious one unless the plant life cycle is carefully examined. By growing plants in the classroom/school garden, children will observe each of the stages of the plant's life cycle. Petals fall from flowers following fertilisation and the ovary noticeably swells.

Cut an apple in half and wipe it with lemon juice to stop browning. Ask learners to examine the cut apple and to look carefully for seeds, flesh and floral remnants. A hand lens will help with this. Ask them to look carefully at the opposite end from the stalk – the sepals and stamens will usually be visible.

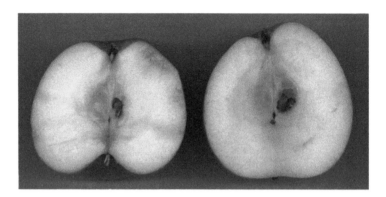

Photo 4.2

© N Souter

ACTIVITY

Seed parts

1   Children can be invited to explore 'an enchanted forest'.

Ask the children to design and sketch the seeds of the plant they 'discovered'. Drawings might show the enlarged details of seed parts.

They should give their plant a common name and possibly a scientific name (if you have shared the required conventions). (You can let them discuss how scientists name new plants when these are found.) They should also explain how the seeds might be dispersed. They can make models of their seeds from junk and craft materials.

Their drawings, explanations and models can be used to formatively assess their understanding of seed dispersal mechanisms.

2   Dandelion seeds store well in airtight containers once they have been properly dried – use these to show learners the shape you are trying to achieve in the model.

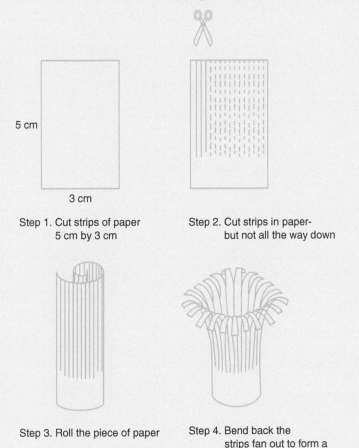

Step 1. Cut strips of paper
5 cm by 3 cm

Step 2. Cut strips in paper-
but not all the way down

Step 3. Roll the piece of paper

Step 4. Bend back the
strips fan out to form a
model dandelion seed

Figure 4.4

Prepare 5x3cm strips of paper (for modelling dandelion seeds). Show learners the basic technique for preparing a model of a dandelion seed. Have a seed-flying contest.

3   Sycamore 'helicopters' are a suitable and familiar stimulus to the pupils. These can be collected in the autumn and will store in airtight containers indefinitely, providing they are well dried. Spinners provide an appropriate model for investigating wind dispersal.

(Continued)

(Continued)

What might affect the flight of a spinner? Give the children this problem. Demonstrate the construction of the spinner. A range of suitable variables may be offered by the pupils: mass of the spinner, length of the wing, angle of the wing, area of wings, height dropped etc. It is important that they are given time to plan their investigation, negotiate with their group and with you as their teacher, and collect some data. The final event can be judged and awards given. Conclude this activity by describing each of the stages that they carried out in terms of investigation design.

4   A plastic drinks bottle and some appropriately-sized polystyrene spheres can be transformed into a model to demonstrate the 'pepper pot' dispersal mechanism of poppies.

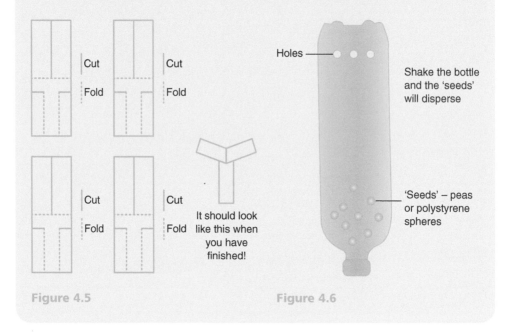

Figure 4.5   Figure 4.6

Water uptake

Botany textbooks frequently indicate that the first stage of germination involves 'the imbibition of water'. Germination involves a series of hormonal and biochemical responses that are triggered by the initial uptake of water.

Photo 4.3

© N Souter

Cress seedlings 48 hours after germination

Children will notice a furry appearance on young roots and more careful examination with a hand lens or digital microscope will reveal root hairs. In nature these permeate the soil and massively increase the surface area for the absorption of water. They will also notice that the shoots have emergent leaves that are turning green and pushing the seed coat away.

Seeds will not usually germinate until they have been removed from the parent plant and from the fruit. In those plants that are dispersed by animals internally (i.e. which pass through the alimentary canal), the soft materials in the fruit are digested while the seed coat protects the embryo from the digestive processes and enzymes (e.g. raspberries, strawberries and tomatoes).

ACTIVITY

Tomato seeds

Compare the germination of fresh tomato seeds, dissected from the fruit that remain surrounded with 'jelly', and seeds from a packet which have the 'jelly' removed. How does the jelly around tomato seeds affect tomato-seed growth? The children should consider the importance of animal dispersal and some interesting ideas will emerge. Does the jelly/tomato pulp affect the germination of the dried seeds?

ACTIVITY

Grass germination

Germinating grass can provide insight into germination and the unusual pattern of growth (grass grows from its base so that the growth point is not removed by grazing).

Make a ball of vermiculite (available from garden shops), contain this within some old tights, and leave a tail that will act as a wick. Support it on a cup or glass and make sure that this wick extends into the water. Once the vermiculite ball has been soaked, use the ball to coat the upper surface with grass seeds.

Photo 4.4

© N Souter

Photo 4.5

Photo 4.6

'Googly eyes' will add to the entertainment while the children observe grass germination and growth. As it grows the grass can be trimmed and 'styled' over several weeks emphasising basal growth.

Vegetative propagation

While flowering, the sexually reproductive life-cycle stage, leads to seed production and resultant genetic diversity: we should note the significance of the way plant growth relies on asexual means, growth that results from the production of new cells and their subsequent enlargement.

Unlike animals, plants continue to grow throughout their lives. (Animals do not normally get bigger once they reach a certain size.) This continual growth in plants happens because they have special cells which form a type of tissue called a meristem. Cells in the meristem divide, grow and develop into different types of specialised plant cells.

Plants have varying lifespans. Some plants (annuals like pansies and marigolds) will live for only one year. In their one-year lifespan they will germinate, grow, flower and produce seeds before dying. Other plants will live much longer – sometimes for hundreds of years. Such perennials include trees and grasses.

Plant propagation by vegetative processes

Plants propagate themselves naturally by vegetative processes. Plants are able to reproduce by forming genetically identical copies of themselves. This reproduction is asexual because it does not need fertilisation to occur and only involves one parent. Asexual reproduction is useful to the plant as it is a rapid method of reproduction that allows it to spread quickly, colonising new habitats. However, all of the offspring are genetically identical and lack variety.

Case 1: Potato plant – the formation of tubers

The potato plant transports excess glucose beneath the ground where it is converted into starch. Underground stems swell to form tubers. These tubers provide the energy needed for growth during the following season. As the food store is used up, the tuber shrinks in size. Once the new shoots emerge from the soil, and leaves form, photosynthesis will begin again, and thus the propagation of the potato continues, season after season.

Case 2: Daffodil – growth from bulbs

The daffodil plant reproduces by forming bulbs. A bulb is an underground food store which provides the energy required for new growth at a later time. Thickened leaves act as food storage organs, providing energy for new root and shoot growth.

Case 3: Spider plants – forming a runner

Spider plants are popular houseplants as they are easy to look after and reproduce successfully. These plants reproduce by forming a side shoot called a stolon, or 'runner', which grows horizontally along the ground. At a short distance from the parent plant, roots may form and embed themselves in the soil. As these roots begin to take up water and nutrients independently, growth of a new plant begins above ground.

Plant propagation by artificial processes

House and garden plants can be propagated artificially by vegetative processes in a variety of ways. Gardeners often exchange plants and move them around their own plots.

Division involves simply splitting plants into smaller clumps (this can be done with a common garden plant, the hosta).

Stem cuttings can be taken from plants. The cut stems can be encouraged to grow new roots. This can be done with such plants as fuchsias and pelargoniums. An African violet will grow a new plant from a single leaf which has had its stalk inserted into potting compost.

The success of these techniques is often helped by carefully controlled environmental conditions as well as plant hormones. These methods are also used to increase the numbers of rare, threatened and endangered plant species.

Vegetative propagation, both natural and artificial, involves the growth of new plants from parts such as the leaves, stems and roots. Since it does not involve sexual reproduction – flowering – the plants that are produced are all genetically identical to the parent plant. Their cells contain exact copies of the genetic material (i.e. the chromosomes in the nucleus and all of their cells are unchanged and they are clones of the parent plant).

Vegetative propagation and differentiation

Vegetative propagation depends on plant cells' ability to change from one type of cell into another. For example, taking a cutting involves some of the cells in the stem changing into root cells. These cells have the potential to specialise into any other plant cell. Remember that the nucleus includes the genetic plan for the whole plant, and that particular genes are being switched on and off so that the cells can grow into particular tissues.

Micropropagation

Plant cells are so good at dividing and differentiating that an individual cell can be grown into a complete new plant. Valuable plants such as orchids can be grown by this process of micropropagation as the clones formed will be disease free.

ACTIVITY

Shoots and roots

1 Learners can grow carrot tops or other root plants (e.g. beetroot, turnips or parsnips) on a saucer on a windowsill. They must make sure that the water is topped up but also that the plants are not flooded above the green parts. When plant cells grow they can then develop into all the other different tissues. Since these cells are genetically identical we are *cloning* the carrot. This property lies at the basis of tissue culture in plants and animals. Stem cell research and therapy are derived from such approaches.

Photo 4.7
© N Souter

Look for the appearance of shoots and roots over a few weeks. Keep a photographic record for display purposes.

2 Collect 33cm lengths of winter twigs (e.g. oak, birch, and horse chestnut) during late winter, and cut the bottom 2–3cm from the stems which helps water uptake (you should always do this with cut flowers at home) and place these into vases. Make sure these do not dry out. Learners compare the bud opening between the different specimens as well as making comparisons with what is happening outdoors. (The cutting may also produce roots over time.)

3 Learners can also grow some plants from cuttings. A few stock plants in school can provide source materials (e.g. spider plants where the plantlet's roots are visible on the parent plant and will rapidly grow when they are placed in damp potting compost; or African violets that can provide source material for taking

(Continued)

(Continued)

leaf cuttings). An essential ingredient, in addition to suitable plant materials and equipment, is patience. Learners will want to see how their plants are getting on, and while their enthusiasm should be encouraged, the plant material should be disturbed as little as possible.

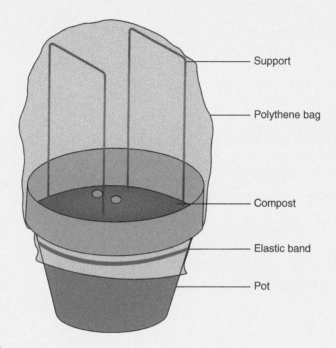

— Support

— Polythene bag

— Compost

— Elastic band

— Pot

Figure 4.7

This simple 'mini greenhouse' can help with germination or propagation experiments.

Sexual reproduction in flowering plants

While we have seen that how seeds work might be obvious to the children, flowers – due to their extraordinary variety – are less obvious. Compare orchids and pansies to grass flowers or willow catkins for example. Look at apple blossom in the spring which will lead to fruit harvest in the autumn. Grass flowers are fairly inconspicuous, as are oak flowers, yet we are all familiar with seeds and hayfever caused by grass plants and the familiar oak seeds, acorns. (Flowers make fruits, fruits contain seeds, and the life cycle continues.)

The key to understanding flowers is one of 'theme and variation'. Flowering plants have colonised all habitats although they are most successful on land. A quarter of a million flowering plants have been classified. We can really only estimate how many more species exist.

Although their diversity is bewildering, there are only four basic floral parts – the sepals, petals, carpel and stamen. Each can be modified to an extraordinary extent, leading to the variety of flowers and pollination mechanisms that exist today.

Figure 4.8

Flower dissection

Flower variety makes the description of their basic parts straightforward, but their exemplification is quite problematic due to the ways in which adaptations have taken place during evolution. It is worthwhile to note the prime purpose of the flower is reproductive and to bring the male cell (pollen) towards the female one (ovule).

A 'typical' flower is made from the four structures mentioned above which are arranged in concentric circles around a short stalk. The sepals and the petals are non-reproductive structures. Sepals – the outermost ring – are frequently green and enclose the developing flower bud. Next come thin, coloured petals to attract insects to the flower. The male part includes the stamens that are responsible for producing male cells, pollen, and the innermost ring; the female part is responsible for producing female cells, the ovules. A 'typical' flower frequently refers to dicotyledonous ones and specimens that show the four rings need careful selection. (We would suggest a hardy garden bush fuchsia as being suitable for class activities.)

Food and oxygen production

All living things depend on energy. Plants absorb energy from sunlight: they need light, water, carbon dioxide, and nutrients to grow.

Plants make food and oxygen. Photosynthesis is arguably the single most important process on the planet today. Radiation from the Sun provides the energy for plants containing chlorophyll to make glucose through the process of photosynthesis. Water is split into its constituent elements (hydrogen and oxygen). Carbon is 'fixed' into sugars and all other organic materials.

During the earliest stage of evolution living things (anaerobic prokaryotes) survived in an atmosphere of methane, ammonia, water vapour and neon, but no oxygen at all.

Photosynthesis provided the evolutionary leap that led to the mass extinction of most other microorganisms since oxygen was toxic to them. Populations of photosynthetic microorganisms then grew, multiplied and gave rise to the accumulation of oxygen in the atmosphere, thereby changing it and in turn enabling the evolution of eukaryotic cells and multicellular organisms. Oxygen's appearance changed the composition of the Earth's crust, its atmosphere and the processes that shaped climate, biodiversity and evolution. Food webs are dependent on producers, and the main ones are plants. Life in its current form could not exist on this planet in the absence of photosynthesis; civilisation and its fate are reliant on the continuing availability of oxygen at present levels in the atmosphere.

ACTIVITY

Gas production

Gas production can be observed in pond weeds. Collect Canadian pondweed (*Elodea*) from local ponds or streams. *Elodea* is also available from local pet shops and aquarium supplies.

Place a jar or tank of the weed in a well-illuminated place (e.g. a windowsill) and bubbles will appear on the leaf surfaces. Cut stems will release gas bubbles when they are inverted in a glass. Small weights (e.g. solder or lead tape) on the end of a shoot will prevent the weed from floating to the water surface. The rate of gas production (oxygen liberation) will increase as the level of illumination increases.

The chemistry of photosynthesis

Photosynthesis involves a complex series of biochemical reactions that involve two distinct operations – one is driven by a light-dependent reaction and splits water to release oxygen, and a second light-independent stage combines hydrogen (H) from water with carbon dioxide (CO_2) to produce carbohydrate (CHO). Note that these reactions run simultaneously inside plant cells in different parts of the chloroplast.

Leaves are designed to absorb light energy from the Sun. In order to do this they provide a large surface area that faces the light and they are thin in order to let light pass through to the lower leaves.

ACTIVITY

Photosynthesis

1   Ask the children to stand beneath a tree in the school grounds and look upwards.

Can they see how the leaves overlap each other and form a mosaic pattern in order to trap as much light as possible? The ideas associated with a mosaic can be developed more fully in mathematical contexts with symmetry and tessellation (in their use in art e.g. Roman flooring and its associated historical significance).

2   Prepare mosaics with pressed leaves.
3   Look more closely at Elodea leaflets with a hand lens and also a digital microscope. Note the shape of the leaf, the internal features (e.g. the leaf edge) and whether any veins are visible. Note how thin the leaf's width is compared to its breadth.

Chlorophyll production

During germination the seed leaves of mustard or cress will turn green as soon as they emerge from the seed into the light. This rapid appearance/synthesis of chlorophyll supports seedling growth and removes the plant's dependence on the seed's food store.

ACTIVITY

Chlorophyll

Prepare two pots with moist compost and sow 25 mustard or cress seeds on the surface. Place one pot on the windowsill and the second one in a totally darkened cupboard or suitable container: you will need to keep the experimental seeds in

(Continued)

(Continued)

complete darkness throughout the week, so no peeking! Compare their growth after that one week. (The plants that have grown in light will be green and healthy, having produced chlorophyll; the seedlings that have grown in the dark will be long and yellow. The seedlings will remain yellow since they have been unable to make chlorophyll. This length increase 'etiolation' is an adaptation that results in extra growth to search for light.)

Energy transfer

Significant energy changes take place in plants. Photosynthesis results from the transformation of light energy into chemical energy. The entire metabolism of plants is enabled as a consequence of photosynthesis, and all the chemical changes that take place involve the transformation of chemicals from one form into another (e.g. the production of starch and cellulose is comparatively straightforward from the glucose that is made by photosynthesis; more complex chemical reactions result in the formation of fats and oils, enzymes and other proteins, as well as chlorophyll). Their metabolism also involves releasing energy from their chemical reserves, and respiration brings this about and additionally involves the transformation of chemical energy into heat energy.

Transport in vascular plants

Photosynthesis depends on water being broken down into its constituent elements – hydrogen and oxygen. In order to carry out photosynthesis efficiently, land plants have developed efficient tissues for transporting water upwards from the soil as well as distributing materials around the entire plant.

The veins in angiosperms include two types of tissue which are concerned with transport. Upward water movement takes place throughout the plant from the roots, ultimately evaporating from the leaves (*xylem*). The movement of sugars, hormones etc. that have been produced by photosynthesis is carried out in the second tissue (*phloem*). These tissues provide an efficient plumbing system that supplies raw materials and effectively transports useful materials around higher plants. Note that most of the lower plants (e.g. mosses) lack transport systems and are thereby restricted to living in damp conditions.

Leaves and stems play significant roles in plants' water relations. These are covered with waterproofing materials (e.g. wax and cork). The purpose of these materials is to prevent water escaping from the plant. Significant adaptations take place in those plants that are adapted to living in dry habitats to ensure that water loss is minimised (e.g. cactus leaves are reduced to spines that prevent grazing and the stem has evolved as the main photosynthetic organ).

ACTIVITY

Water transport

1   Place a pot plant inside a transparent plastic bag and look for the appearance of condensation after an hour or two.

2   Look carefully at the upper and lower surfaces of leaves. Use a digital microscope to look for pores, stoma that allow water vapour to escape and carbon dioxide to enter leaves.

3   Take particular care to avoid spillages and scalds while plunging a leaf with tongs or tweezers under the surface of water just off the boil in a Pyrex container. Look for the appearance of gas bubbles on the lower leaf surface in particular. The expanding air is forced out of the leaves and through their pores. Children might be asked to speculate what is happening and to infer the presence of those microscopic pores.

4   Compare the vein patterns that appear in different leaves that the children can collect in the school grounds. Note the differences that appear in monocots (parallel veins) and dicots (branching veins).

5   Collect/prepare leaf skeletons.

6   Experiment with cut flowers, leaves, or celery placed in food dyes. Cut the bottom 2–3cm from the stems which helps water uptake. Place the cutting into water to which a few drops of food dye have been added. Children enjoy seeing for example white carnations turn blue or celery turning red, as well as gaining a secure understanding of water transport in plants.

Photo 4.8

© N Souter

Responding to the environment

Directional growth

Learners may notice seedlings growing towards the window. Such directional responses are the result of growth in relation to specific environmental stimuli. Photosynthetic tissues grow towards light and away from gravity, whereas root tissues will grow downwards.

ACTIVITY

Directional growth

1   Ask the children to design and carry out an investigation to find an answer to this problem: Which way up? Does it matter what way seeds are planted?
2   Prepare pots with moist compost and sow 25 mustard/cress seeds on the surface. Place individual pots in different conditions – in full light; inside a cardboard box with a window cut on one side; inside a cardboard box with no light at all. The variable here is the position of the light. Compare growth after a week noting any directional growth and etiolation (see earlier).
3   Grow bean plants in flower pots for around two weeks until they are 12–15cm high. Record the plants' growth on a chart. A photographic record will help with this. Turn the pots on their sides and allow the plants to continue to grow. Continue recording their growth.

Sensitivity – do plants know?

Plants are remarkably sensitive to light, gravity, and a range of other environmental stimuli. We have seen how light and gravity affect directional growth, however light can provide additional, more subtle, cues in response to day length. Seasonal responses are most dramatic where the latitude results in significant changes in day length. Flowering and leaf fall are the most commonly observed plant responses in relation to changing day length.

Plants are able to detect chemicals in the environment that for example bring about directional growth towards certain materials and away from others. Many release chemicals which prevent growth by competitors. Ethylene gas is used to help fruit ripening in an industrial scale. Most are responsive to touch and respond to the proximity of other plants, thus regulating their growth.

While many children find plants 'boring', they provide fascinating responses to the environment.

ACTIVITY

Polytrichum moss

Collect some *Polytrichum* moss (pronounced 'Pol-it-rick-um'). *Polytrichum* has had various uses around its global distribution. It has been used as a decorative material on Maori cloaks, taken as an infusion to combat gallstones, used to make brooms and brushes, and plaited into various items (e.g. a basket found in the remains of a Roman fort in Newstead, England).

Photo 4.9
© N Souter

Photo 4.10
© N Souter

Polytrichum moss grows in clumps in damp areas, particularly woodland. Collect some of this and keep it in a sandwich bag to prevent it drying out before the lesson.

1   Leave a few strands on a table top for 10 minutes and compare the appearance with the fresh moss. (The leaflets should start to curl towards the stem as they dry out. This is a simple water conservation technique that is used by the moss.)

Ask the children to suggest how to restore its former appearance. Plunging it into water allows this to happen. This impressive and quick reversal indicates a surprisingly rapid response by plants.

2   Children may be familiar with the Venus fly trap, an insectivorous plant that uses a range of mechanisms – scent, colour, enzymes and physical pressure – to trap and digest its prey, and overcome the shortage of nitrogen that it experiences

(Continued)

(Continued)

in its acidic habitat. No less dramatic, and easier to grow, *Mimosa pudica* (the False Acacia) is sensitive to touch and snaps its leaves shut when they are touched. The most important moment is probably when the children express surprise and delight when the leaves collapse and they can then go on to explore the responses, recovery times etc.

Before

Photo 4.11

© N Souter

After

Photo 4.12

© N Souter

Adaptation

Living things are adapted to the environment in which they live. Adaptation is a central part of the theory of evolution and describes the process by which plants survive in particular conditions and the genetic changes that lead to variation in the population and ultimately the appearance of new species.

Since all living things are the product of their evolution it follows that they all include particular adaptations. Similar strategies are frequently adopted by unrelated species to overcome specific environmental problems and challenges.

The desert

Low rainfall, rapid drainage, little shade, and strong winds provide difficult conditions that must be overcome by plants that live in the desert.

Many will store water in their stems or leaves, while others will have no leaves at all or these will have evolved into spines which prevent grazing. They are frequently covered with a thick wax surface that prevents water loss. Deep, fibrous rooting systems absorb water, while others evade drought altogether by surviving as seeds and progressing through their life cycle rapidly following heavy periods of rain. Flowers will often only open at night.

Freshwater

Freshwater plants are frequently filled with air spaces which provide buoyancy and keep the plant at the water surface in maximum light conditions. Their leaves and stems are flexible so that they avoid being damaged by water currents. The transport system is significantly reduced since they are able to obtain carbon dioxide and water, needed for photosynthesis, directly from the environment and then release the waste oxygen into the environment. The roots only play a role in anchorage. Water plant seeds are frequently adapted to float as the dispersal mechanism.

Avoiding predation

Since plants are producers and are found at the start of food chains, adaptations exist to avoid grazing by predators. They sometimes achieve this by including

Photo 4.13 *Nettle leaf*

© N Souter

chemicals that make them taste unpleasant. Seeds often release poisons (e.g. cyanide) to prevent predation. Others frequently adopt strategies for warding off predators (e.g. stings and thorns).

The stinging hairs on the edge of a nettle leaf inject histamine and other chemicals into predators.

Safety

Hygienic practice should always be practised when working with living materials. Plants and all their parts, including seeds, soils and potting media and equipment, should all be regarded as potential contaminants, and handwashing should be carried out at the conclusion of every activity.

Risk assessments must be maintained to ensure that poisonous plants, those that promote allergic responses, and those that might have been treated with poisons have been taken into account.

Eating during these practicals must be forbidden to ensure that the ingestion of poisons or allergens is prevented. Note that this needs to include food plants found in the wild.

Summary

Plants provide exciting and uncontroversial learning opportunities for children. They are capable of being used to illustrate the full range of vital processes in reliable and consistent ways. They are such intriguing organisms that it is paradoxical that plant science is seen as being a soft option; inappropriate gender stereotypes are associated with their study. The most superficial programme that introduces the adaptation of plant parts and their reproduction will inevitably lead to learners being impressed with the extraordinary theme and variation shown for example in leaves, flowers, stems and roots. Equally impressive are the many ways society has used plants for food, clothing, construction, energy, medicine etc. Plants' capacity to respond to the environment is notable. Their asexual and sexual reproductive processes provide opportunities for meaningful learning experiences, as well as setting the scene for further study on genetics, plant breeding, and genetic modification.

REFLECTION POINTS

- Consider the contribution of fossil fuels to global warming.
- Consider the value of using plants for teaching aspects of life science.
- Use a calculator to compare the masses of the smallest and largest seeds.
- Make a mind map to summarise propagation and plants. Include ideas about sexual reproduction and asexual reproduction.
- A significant learning issue surrounds the misconception that plants gain mass from the soil rather than gaining mass through photosynthesis.
- Identify suitable local sites, collect and preserve plant material for classroom investigations and displays.

The origin of this misconception might relate to use of the term 'plant food' when adding fertiliser to replenish mineral nutrients in cultivated plants. While these nutrients (mineral salts) are essential for many of the chemical reactions that take place inside plants, they are not an energy source. Plant growth results from the biochemical processing that takes place in light and combines, or 'fixes', atmospheric carbon dioxide with water from the soil to provide organic chemicals. The mass of an oak tree, the production of cereal and all plant growth depend on such carbon fixation. Which teaching and learning strategies might you adopt to overtake such misconceptions?

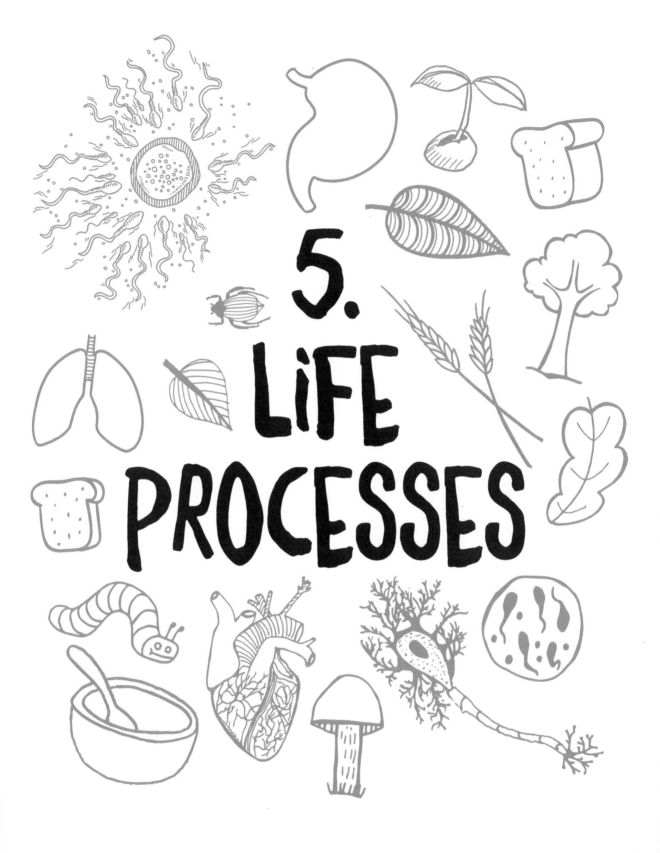

5.
LiFE
PROCESSES

Learning Objectives

Having read this chapter you should be able to:

- identify the features shared by living things
- organise investigative and experimental activities in respiration, obtaining energy, food, support and movement, the circulatory system and the human life cycle
- distinguish between respiration and breathing
- describe aspects of the mechanical and chemical breakdown of food
- explain the roles of muscles, joints and associated structures in movement
- outline key structures and functions of the components of the circulatory system
- provide the framework of human reproduction

Overview

'An amoeba can do everything an elephant can do' the professor exclaimed!

Students debated the meaning of this assertion in terms of basic life processes. Both organisms are able to release energy (1) from food by respiration, they are capable of growth (2) by producing new cells and by exchanging genetic material during sexual reproduction (3). Each organism is capable of detecting the environment (4) with its senses, responding to it by moving (5), and obtaining food (6). Undergraduate humour took over when they compared waste removal and excretion (7) by the selected organisms.

These seven vital processes can be used to distinguish living from non-living materials. They might need some refinement while considering specific cases (e.g. plants' 'senses' are able to 'detect' light and day length, water, minerals and other chemicals, and gravity, but can only 'move' by growing towards or away from the stimulus). These processes might previously have been of philosophical interest only, however the search for life in other parts of the solar system, and beyond, opens the discussion such that it becomes one of particular note and topical interest.

The principal focus in this chapter will be turned towards human physiology which has implications towards adopting healthy lifestyles as well as making appropriate life choices. Note that the chapter will include reference to other animals for comparative purposes; plants obtaining their energy by photosynthesis is covered in Chapter 4 and the senses are included in Chapter 7 alongside animal behaviour.

Each of these vital processes involves complex chemical control and coordination, emphasising the extent of the nature of cell biology and of physiological systems in general. Since living things share these characteristics it follows that they are all extraordinarily sophisticated: whether they exist at the microscopic level (e.g. amoeba) or as leviathan land animals such as the elephant they are specialised, adapted to their environment, and are capable – irrespective of their size, type or complexity – of coordination and control. These ideas lie at the heart of our fundamental biological theory, namely evolution.

Control implies balance. Metabolism inevitably leads to rising and falling levels of nutrients and waste products inside cells and the entire organism. Without control accumulation/depletion will have toxic and ultimately fatal outcomes. Although we do not suggest that balance in physiological systems (homeostasis) features in a programme of study, it is an essential feature of understanding the underlying principles. Coordination is consequently required between the various systems so that internal senses might detect falling nutrient levels that leads in turn to using other senses to detect prey, moving to capture it, and subsequently breaking the food down into nutrients that will redress the imbalance. The useful products of digestion are circulated around the body; waste is removed.

Cells are often aggregated into tissues, tissues into organs, and organs into organ systems. In the human body, systems carry out the key functions of respiration, digestion, elimination of waste, and temperature control. The circulatory system takes material needed by cells to all parts of the body, and removes soluble waste to the urinary system.

Children's awareness of themselves, what is inside their bodies and how they work, will be naïve and inventive.

ACTIVITY

Body awareness

Find out learners' initial ideas and prior knowledge through a combination of draw and write and think-pair-share on a body outline.

Respiration

Respiration refers to the biochemical processes that release energy within cells and organisms. Respiration is ubiquitous in nature. In environmental terms respiration drives one part of the carbon cycle, but in physiological terms it is concerned with the production of ATP which is the biochemical 'currency' that is needed to drive all cellular processes forward. Respiration provides the energy for growth, repair, excretion, synthesis – everything.

Energy release is controlled, as are all physiological processes, by enzymes and can take place in the absence (anaerobic, ancient in evolutionary terms, inefficient, low

ATP production) of oxygen and in its presence (aerobic, more recent in evolutionary terms, efficient, high ATP production).

We can infer some of the required complexities of cellular respiration in multicellular organisms – obtaining food; digesting it into a usable, soluble form; obtaining oxygen and ensuring that a continuous supply is provided to all parts of the cell/organism; removal of the potentially toxic acid waste that is formed when carbon dioxide dissolves in water. The 'simplest' organism includes complex adaptations to ensure its survival.

Chemical equation for aerobic respiration

glucose + oxygen → carbon dioxide + water (+ energy)

Respiration is controlled by enzymes. The energy contained within food (e.g. glucose) is released in the presence of oxygen as ATP, useful cellular energy. Carbon dioxide forms acidic waste that must be removed.

Energy transformations in nature

The first law of thermodynamics states that when energy is transformed from one type into another the total energy remains unchanged.

- Vital processes involve energy changes.
- Respiration releases energy in the form of ATP and this allows living things to do their work.
- Light, from the Sun, is the principal source of energy for food chains.
- Photosynthesis is the biochemical sequence that transforms light energy into the plant biomass.
- Metabolism in plants and their consumers converts those chemicals from one form into another as well as making heat.
- Movement is brought about by respiration in muscle tissues which act on the skeleton for locomotion; chemical energy in food is converted into that movement (and heat).
- The environment is rich with sounds that are made by animals courting, defending their territory etc. which involves converting chemical energy into sound (again with some heat production).
- Animal electricity is dramatically demonstrated in certain species (e.g. the electric eel which converts chemical energy into electricity to stun its prey).
- Nervous impulses involve transforming chemical energy into electricity (plus heat).

- Several organisms are able to convert chemical energy into light (e.g. glow-worms and bioluminescent microorganisms), remembering of course that significant energy losses take place due to heat production.

Gas exchange

Respiratory gases move across membranes from high to low concentrations. This is straightforward in single-celled organisms since they are in direct contact with their environment. Multi-cellular organisms have developed a variety of organ structures to increase the efficiency of gas exchange – oxygen uptake and carbon dioxide release – and this relies on having moist surfaces which permit the respiratory gases to dissolve, having large surface areas over which gas exchange can take place, and developing effective circulation to transport the dissolved gases.

ACTIVITY

Gas exchange

1 Examine earthworms to note their large surface area in comparison to their volume (this can be modelled with a sheet of photocopy paper rolled into a cylinder), as well as observing their moist surfaces.
2 Collect some garden snails and observe the movements of the breathing pore.
3 Observe various stages of the frog life cycle. In its earliest stages a tadpole has external gills and as it grows larger the gills are found inside the tadpole. When it metamorphoses into a froglet dramatic changes take place with the development of lungs and the retention of a damp skin so that gas exchange takes place in two ways in adult frogs – breathing via lungs and dermal gas exchange.

Breathing

Breathing involves inhaling and exhaling air and it enables gas exchange to take place inside the lungs. We really don't think about breathing and yet our lives depend on continuous, unobstructed access to suitably oxygenated air.

Breathing is brought about by ventilatory movements all around the lungs which take up the greatest amount of space inside the thoracic cavity. They are surrounded and protected by the shoulder girdle at the top, the rib cage around the perimeter, and the diaphragm at the floor. They have a rich blood supply and receive blood from the right side of the heart and return it directly to its left side. Moist surfaces, a huge surface area and a rich blood supply in the air sacs ensure that oxygen is able to enter the blood and that effective carbon dioxide removal takes place.

ACTIVITY

Breathing

1   Ask the children to sit in pairs facing each other. (Children with asthmatic and other chronic respiratory conditions should observe these activities only.) Ask them to carefully watch each other while they breathe in, hold their breath for three seconds, and then breathe out. Repeat this two or three times so that they become used to the activity.

Ask them to touch each others' shoulders and watch what happens when they breathe in and out. (Their shoulders should rise during inhalation and fall during exhalation.)

Ask them to place their index fingers against their ribs while they breathe in and out. (They may be able to detect those ventilatory movements that are brought about by the intercostal muscles.)

As them to place their fingers against their solar plexus and feel, and watch what happens when the deepest breaths are taken. (The diaphragm pushes down on the belly causing it to move outwards.)

2   Children may be familiar with 'spare ribs' and able to explain that meat lies between the bones; these are the intercostal muscles. Collect a small sample of lamb ribs from the butcher and examine the intercostal muscles. The relationship between muscles, movement and meat is rarely obvious to children.

3   Construct model breathing systems with a plastic water bottle, 30x30cm plastic (shopping bag plastic will do), one small balloon, a rubber band or masking

Figure 5.1

tape. The children should construct their own models and then experiment with them, noting what happens when they move the handle in and out (thereby simulating the movement of the diaphragm).

Remove the end of the bottle as shown and push the end of the small balloon through the mouth of the bottle. Then stretch the neck of the balloon over the neck of the bottle. Fold the plastic square twice and twist the folded corner and tape it to provide a handle. Secure the plastic to the bottom of the bottle.

4   Children should measure their chest movements during breathing. This can be achieved simply with string or linked to numeracy with measuring tapes; the children should notice that the circumference of the thorax (their chest) increases during inhalation and falls during exhalation. Differences can be measured and recorded in class displays.

5   Artificial resuscitation and CPR might be included during the upper stages and can involve appropriate interest groups (e.g. the Red Cross).

6   Research childhood respiratory conditions (e.g. viral respiratory tract infections such as the common cold, influenza (the flu), croup, bronchiolitis or pneumonia, or allergic conditions such as asthma).

Obtaining energy

All living things need a source of chemical energy (food) for growth and repair as well as reproduction, movement and every other vital process. Some animals eat plants, and some eat other animals, but all animals ultimately depend on green plants for food. Green plants can make food, animals cannot. The distinction between autotrophic and heterotrophic lifestyles and their impact on food webs has been explored in earlier chapters. The position of an organism within a food web is able to inform us about aspects of its lifestyle, its ecological niche, and the associated predator/prey relationships.

Humans make use of animals and plants to provide food; they are omnivorous. Our digestive system is designed to process the widest range of food materials. An animal's entire digestive system is adapted to its diet and generalisations appear in relation to herbivorous, carnivorous, and omnivorous lifestyles.

The principal purpose of digestion is to convert insoluble food materials into soluble ones that are able to be transported around organisms and inside cells. This is achieved in two ways – through the mechanical and chemical breakdown of the food. Mechanical breakdown frequently involves teeth or structures that cut food into smaller chunks, and it also involves muscular movement inside the alimentary canal that will churn food and turn it into pulp, as well as mixing it with digestive enzymes which are responsible for the chemical breakdown and the final conversion from insoluble into soluble materials.

Digestion and food breakdown

The physical breakdown of food – chopping food

In multicellular organisms the first part of processing food (i.e. ingestion) often involves cutting the food into smaller parts so it can smoothly pass through the alimentary canal.

ACTIVITY

Chopping food

1 Bite into a polystyrene cup and look at the impression that is left from the front teeth.

2 Watch a partner take a bite from a slice of bread. Do they use the same teeth to cut the bread as they do to chew the bread?

3 Observe and record a family pet eating – cats and dogs will gulp their food and occasionally cut it into smaller pieces with the teeth at the back of the jaw. Small mammals (e.g. hamsters and gerbils) gnaw their dried food. Video evidence will help here and allow comparisons to be made between the teeth used, the dentition, and the feeding style.

4 Visit a local farm and watch the way cattle, sheep or goats feed. Note the use they make of their tongue and lower front teeth to help cut the grass. Note also the use that they make of their rear teeth to shred the food. It may also be possible to observe animals 'chewing the cud'. Video records should be established in order to make further comparisons and stimulate further enquiry.

5 Compare the dentition of humans with that of carnivores (e.g. cats and dogs) and herbivores (e.g. cattle and sheep) on museum specimens. Note the scissoring action of the rear teeth in the carnivores that is brought about by the hinged action of the jaw. Note also the grating properties of the herbivore skull when their jaws move from side to side.

6 Simulate a dental inspection by inviting learners to count the numbers of teeth that they can find in the upper and lower jaws. Developmental issues will emerge and the loss of deciduous teeth can be highlighted. Pupils may be able to supply collections of their milk teeth and these can be examined to note the hard enamel surface as well as making comparisons with their adult teeth, stressing that they have these for life and the importance of dental care.

7   Integrate work in this area with work related to dental care and hygiene, frequent brushing and regular dental visits, as well as noting the importance of fluoride.

8   One important aspect of dental care includes removing plaque-forming bacteria which release acids that in turn result in dental decay. Similarly, concern is expressed regarding the ingestion of fizzy drinks which also contain acids. Dried egg shells can be used to demonstrate the destructive effects of weak acids. Egg shells are calcareous in nature and have chemical similarities to teeth, and are therefore useful for comparative purposes. Immerse egg shells in a variety of liquids – water, a selection of fizzy drinks, vinegar – and record the changes. Photographic records can provide stark insights into this experiment.

9   Collect some garden snails and paint a trail of flour and water on a Perspex sheet. Once the snails are actively feeding support the Perspex sheet vertically and watch them feeding from the other side. Note the tongue rasping the food mixture from the surface of the Perspex. You might also see ripples along the snail's foot as it moves along; this can be compared to peristalsis (see later).

10 Watch caterpillars feeding. This might be possible from specimens collected in the school grounds, from livestock purchased from accredited suppliers, or by a visit to a butterfly farm. Closely observe the cutting action of the insect's mouthparts and compare these to the adult butterflies feeding on nectar. Butterfly mouthparts can be modelled with party blowers. Invite the children to speculate about the differences in feeding mechanism and also consider the dramatic change that has taken place during metamorphosis.

Photo 5.1

© N Souter

The physical breakdown of food – moving food

The problem of moving food through the alimentary canal of a multicellular organism is overcome by lubricating the food/enzyme mixture and squeezing it along the entire length by muscular activity. In the human digestive system lubricating liquid is provided in the mouth (salivary glands), in the stomach (gastric fluid including acid), from the liver (bile) and the intestinal walls. Clearly a lot of fluid is added throughout the digestive processes that must then be reabsorbed to prevent excessive fluid loss – this is the role of the large intestine. The alimentary canal has circular muscles and longitudinal muscles which work together to squeeze the food along the length.

ACTIVITY

Moving food

1   Encourage children to listen to the noises that food makes as it travels along the alimentary canal. Toy stethoscopes are surprisingly efficient at picking up noises that are made in the belly as food is squeezed along its journey. Simple microphones attached to a laptop provide suitably entertaining results.

2   Earthworm movement resembles the way in which the intestines move food. Take care to avoid excessive handling of the earthworms, as well as ensuring that they spend the shortest time removed from appropriate containers. Observe the movement as they traverse a sheet of paper. Note that they become progressively straighter, becoming longer and thinner, anchor and then become shorter and fatter. Successive movements along different parts of the earthworm result in forward motion; longitudinal muscles relax while circular muscles contract as it squeezes forward, and the reverse happens when each segment becomes shorter and fatter. Similar peristaltic locomotion is carried out in blowfly maggots and in each case you can often see food waste (faeces) appearing at the rear end. The rhythmic waves that can be observed in snails as they glide over Perspex can also help children understand peristalsis.

The physical breakdown of food – mixing food

A significant part of the alimentary canal is concerned with mixing food with fluids to ensure it is adequately lubricated, and with acids and alkalis that assist digestion and regulate microbial populations, as well as providing optimum pH for the enzymes which progressively convert insoluble food into dissolved nutrients.

Some materials cannot be digested and yet these are an important component of the mass that travels through the alimentary canal since it provides bulk, roughage and material that is expelled from the body as waste.

Food is mixed with saliva and starch-digesting amylase in the mouth. It is mixed with strong hydrochloric acid in the stomach which assists gastric enzymes to digest protein while providing a sterilising barrier that significantly reduces populations of bacteria contained within the food. Muscles in the stomach (think of this as the meat that is eaten in tripe) work like a tumble dryer in churning, squeezing and squashing the food together with the acid and the enzymes, turning it into a fine paste.

The liver and pancreas squirt fluid into the small intestine for a variety of reasons – to neutralise the mixture, to remove waste from the liver (e.g. the bile pigments that provide faeces with the characteristic colour) – as well as adding digestive enzymes.

The small intestine continues to add digestive enzymes so that carbohydrate is broken down into simple, soluble sugars, protein is broken down into soluble amino acids, and fats are broken down into glycerol and fatty acids. Mixing continues in the small intestine to ensure effective digestion takes place and that the food/enzyme/liquid mixture is exposed to the variety of bacteria that contribute significantly to making nutrients, vitamins and minerals available to the host organism. The main purpose of the large intestine is to reabsorb water, prevent dehydration, and avoid diarrhoea.

Since protein is comparatively easy to digest carnivores tend to have a short alimentary canal; plant material is difficult to digest. Complex polysaccharides cannot normally be digested by multicellular organisms and so they frequently rely on microorganism partners to digest these materials. Cattle and other ruminants have large populations of cellulose-digesting microorganisms in one of their stomachs, so that when they 'chew the cud' they are regurgitating partially digested grass which will be further digested in the remaining stomachs and lengthy intestine system. Such partnerships are not restricted to vertebrates and the success of termites depends on the microorganisms that colonise their alimentary canals to digest cellulose. These microorganisms are of significant interest to those scientists who are exploring the potential uses of biofuels.

ACTIVITY

Mixing food

1 Place orange juice into a ziplock bag with a few bite-size chunks of toast and a handful of cornflakes with some milk to simulate a typical breakfast. Seal the bag and ask the children to squish and squash it for a few minutes. The food materials should quickly break down into a pulpy mess, simulating the activity inside the stomach. Children may also note that the mixture has become warm (this would help enzymes) and resembles 'sick'!

2 There are several ways of continuing this simulation so that the full process of food being moved through the alimentary canal can be simulated, and this often involves the use of tights, food dyes, water reabsorption, and the production of 'poo'.

The chemical breakdown of food

The mechanical breakdown of food releases few of the nutrients that are contained within it. Digestive enzymes that are released at various points in the alimentary canal are responsible for the progressive conversion of complex, insoluble materials into those soluble ones which can be absorbed across the intestinal walls and transported in the circulatory systems to the rest of the body.

Digestive enzymes are released at various points in the alimentary canal; each enzyme is very specific in the reaction that it controls, and their products are circulated in either the blood or the lymphatic system.

Table 5.1

Enzyme	Reaction	Location	Product circulation
Amylase	Starch to sugars	Salivary glands, pancreas, small intestine	In blood
Protease	Protein to amino acids	Stomach, pancreas, small intestine	In blood
Lipase	Lipids to fatty acids and glycerol	Pancreas, small intestine	In lymph

ACTIVITY

Chemical breakdown of food

Ask the children to chew a small piece of bread, no more than a quarter of a standard slice, but not to swallow it. Initial physical breakdown and mixing with saliva will be followed by the taste becoming progressively sweeter, as the starch is converted into maltose sugar by saliva which contains an amylase.

Food

Food staples and their sources

Staple foods are those which make up the dominant portion of the diet and provide the majority of the energy required. A variety of different staples exist around the world and these have significant relationships with the evolution of human society and the changes from a hunter-gatherer to an agrarian and ultimately urban society.

Staple foods almost entirely come from plant material. Several staples come from the grass family and are collectively referred to as cereals (e.g. wheat, maize, barley, oats, rice, and rye). 'Pulses' are dried seeds from members of the pea family and include peas, beans and lentils. Underground plant materials (e.g. roots and tubers) provide staples such as potatoes, yams and cassava. (You are possibly unaware of having eaten cassava yet in its processed form it is known as tapioca.)

ACTIVITY

Staple foods

An exploration of staple foods can present opportunities for celebrating cultural diversity, as well as noting that ethnic groups' diets are determined by climate and the success of crop plants in each area. Consider the use, distribution and cultural significance of cereal-based foods (e.g. bread, roti, pasta and rice).

Food, nutrition and diet

Food (material that is ingested in solid or liquid form) contains nutrients in particular proportions.

Nutrition is the process of obtaining food, carbohydrate, protein, fat, vitamins, minerals and water. Dietary fibre may be considered a nutrient since it ensures the movement of food through the alimentary canal.

Diet refers to the usual food and drink taken by an animal. Feeding relationships are described within food webs.

Human diet is a complex topic of contemporary interest relating to health and well-being. Significant concerns are expressed in relation to the impact of poor diet on for example the obesity epidemic, cardiac disease, and dental decay. Significant political and commercial interest is expressed in relation to nations' diets; regulation of food intake is frequently used for weight-loss purposes.

The target is to maintain good health and this can be achieved, in part, by maintaining a balanced diet. Healthy body mass can be calculated on the basis of factors such as height, gender and age. A healthy diet is judged by maintaining a healthy body mass. A balanced diet additionally needs a wide variety and correct proportions of different foods. Approximate, appropriate proportions are indicated on the 'eatwell plate' (see Figure 5.2).

Malnutrition occurs when the diet is not balanced and does not contain the correct proportions of nutrients. Although it is frequently exemplified through undernutrition leading to starvation, it is important to note that overnutrition also represents a significant aspect of malnutrition. Too often the focus is on dietary shortcomings (e.g. vitamin deficiency), and it is equally important to acknowledge the significance of overnutrition on general health.

Fruit and vegetables

Bread, rice, potatoes, pasta
and other starchy foods

Meat, fish, eggs, beans
and other non-dairy
sources of protein

**Food and drinks
high in fat and/or sugar**

Milk and dairy foods

Figure 5.2

Crown Copyright, courtesy of the Food Standards Agency (2010)

ACTIVITY

Diet

1 Discuss the significance of what is shown in the eatwell plate; discuss largest portions and smallest ones; compare proportions; speculate the outcomes of having an excessive dependency on each of the five food groupings contained within the eatwell plate.

2 Investigate the proportions of nutrients indicated on food labels; show these as proportions on paper plates.

3 Research groups should each investigate one of the following dietary disorders – kwashiorkor, obesity, marasmus, rickets, or anorexia nervosa. Provide structured questions to be answered within each topic. Share key findings in a plenary.

Food additives

Materials are added to food to help preserve it and enhance its flavour. Pickled materials are soaked in vinegar, an acid which prevents bacterial growth and food spoiling; salting meat and fish also prevents bacterial growth and removes a lot of water from the preserved food (e.g. bacon).

Preserved and processed foods now contain a wide range of flavour-enhancing and preservative chemicals that are approved by legislation. Some additives occur naturally and others are produced by the chemical industry. They can be used for colouring, sweetening, preventing oxidation, preserving, modifying texture, and as flavouring.

Food additives, essential in food preparation e.g. seasoning, can be controversial and the physiological effects on individuals can be variable; some additives can be harmful at certain levels in particular individuals.

ACTIVITY

Food additives

1   Research one of the key uses of food additives (colouring, sweetening, preventing oxidation, preserving, modifying texture, and flavouring).
2   Audit food labels to identify the occurrence of food additives.
3   Separate food dyes by using chromatography (e.g. from Smarties).

Support and movement

Support and movement in animals are brought about by the coordinated actions of the muscular and skeletal systems. Finding out about their own muscles and bones generates lots of interest for children while they gain insights about how their bodies work. Generalisations about support and movement can be made by observing other animals, both vertebrates and invertebrates.

The skeleton

Skeletons generally have four purposes in animals: they define the animal's shape; they support the organism; they provide muscle attachment which facilitates movement; and they protect internal structures. Additionally, vertebrate skeletons are involved in blood cell production in the bone marrow.

Children are frequently surprised to learn that they have bones inside their body. That bones are more than inert materials only becomes apparent when broken bones show the capacity to repair.

ACTIVITY

Bones

1   A collection of bones provides a useful introductory resource to establish the skeleton's purpose in terms of support, movement and protection. Such first-hand, direct contact provides irreplaceable experiences.

2   Encourage the children to feel the bones inside their body. Make frequent reference to the fact that they bring their skeleton to school every day.

3   Try to make a rag doll stand up. Insert 'bones' (lolly sticks) to create a rigid skeleton that allows it to stand up.

4   Play the game 'Simon Says', naming various bones appropriate to the children's level of understanding.

5   Use a wooden artist's mannequin, stringed puppet, or model skeleton to show joint movement – can the children replicate these movements?

6   Use pipe cleaners and drinking straws to model various skeletal parts (e.g. hands and arms).

7   Explore the variety of joints that occur in the body – hinge joints at the elbow and knee in one direction only; ball and socket joints in the shoulder and hip allowing rotation; and fixed joints (e.g. hip girdle and skull) where the bones are fused together. Relate these to the principles for levers (see Chapter 18).

8   Assemble a skeleton model using paper skeleton parts and paper fasteners.

9   Examine x-rays of bones (countless images are available on the Internet).

10   Visit a local museum that displays animal skeletons and review previous learning in terms of the function of the skeleton.

Alternative skeletons

The familiar internal vertebrate skeleton, with its simple organisation of a central column to which four limbs and a protective 'brainbox' are attached, represents the minority position inside the animal kingdom – external skeletons and the absence of a skeleton altogether occur in many more animal types.

Exoskeletons

Hard-bodied invertebrates, especially the arthropods (e.g. insects, spiders, crabs, woodlice, centipedes, and millipedes), all possess a hard external skeleton to which jointed limbs are attached. Their internal organs are suspended inside body fluids.

ACTIVITY

Body types

1 Collect invertebrates to observe the diversity of body types contained within the arthropods, noting the organisation of the jointed limbs and recording these photographically.
2 Design model beetles from recycled egg boxes, pipe cleaners, bendy straws, and other craft materials.

No skeletons!

Earthworms and slugs, as well as many other types of invertebrate, do not possess a hard skeleton, yet are very well suited to their environment. Earthworms possess two sets of muscles – short circular muscles surround each segment, and longitudinal muscles run from segment to segment for the length of the earthworm. Movement is only possible as a result of squeezing the body fluids. Anchorage in soil tunnels is achieved by the widening of the segments, as well as gripping the tunnel with bristles called setae.

ACTIVITY

Earthworm movement

Earthworms can be collected from school grounds, home garden or from fishing tackle suppliers.

1 Observe earthworm movement. It may help to make a video recording to permit repeated observations of their movement (thereby reducing the stress on the animal) and allowing it to be observed more carefully at a reduced playback speed. Keep the time the animal is under observation to a minimum to prevent distress and desiccation of its skin.
2 Place an earthworm on the centre of a piece of recycled photocopy paper parallel to the long side. Roll the sheet along its long axis into a cylinder 4–5cm in diameter. Listen carefully to the scratching sound made by the moving earthworm while the setae move over the paper. Keep the time the animal is under observation to a minimum to prevent distress and desiccation of its skin.
3 Gently stroke the undersurface of an earthworm to detect the bristles (setae).
4 Simulate earthworm movement with a partially inflated long balloon.

The way we move

Our skeletons provide a combination of bones which are connected by ligaments, and articulate with each other at joints. Movement is brought about by the contraction of muscles, which are attached by tendons around the joint. The relationship between meat and animals' muscles is not always obvious to children, and direct observation of chicken legs or wings for example can reduce this misunderstanding, as well as permitting observation of the component parts.

The two big ideas here are that movement is brought about by muscular contraction and that muscles always work in pairs.

ACTIVITY

Movement

1   Activities in the gym can be used to help identify the muscles which we use to stand, to push, to raise our knees, to touch the sky. While it is more difficult to identify the muscles which reverse each of these, the principle of controlled movement can help here.

2   Use the body to move in different ways (e.g. crawling, rolling, hopping, skipping, etc.).

3   Ask the children to hold one arm out horizontally, to feel the biceps contract when their arm bends, and the triceps contract while their arm straightens.

4   Can they stand or lie in a completely rigid position? Can they make their arms, legs, fingers, and body floppy, and then strong?

5   Compare their body bending to the joints on the skeleton.

6   Opportunities may arise to discuss both posture and disability.

7   Dissect a chicken leg or wing for example, and observe the joints, cartilage, ligaments, tendons, muscles/meat in their opposing pairs, the fibrous nature of the meat, and the bones to which they are attached. Good scissors and a sharp kitchen knife are required here and need appropriate control and safety measures.

8   Model arms can be constructed (see Figure 5.3).

The circulatory system

The circulatory system includes the *heart, blood and blood vessels*. Its purpose as a delivery system is to transport respiratory gases, nutrients, hormones and blood cells, but it is also involved in physiologically balancing the entire body's temperature, pH etc., as well as playing the central role in combating infection within the immune response.

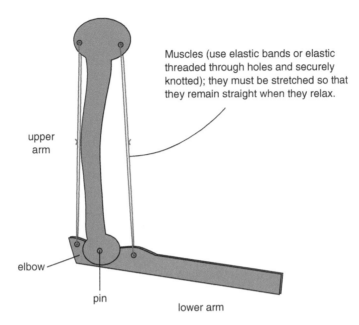

upper
arm

Muscles (use elastic bands or elastic
threaded through holes and securely
knotted); they must be stretched so that
they remain straight when they relax.

elbow

pin

lower arm

Figure 5.3

In humans the circulatory system encloses the blood inside other organs – the heart, the arteries, and the veins. The fist-sized heart lies slightly left of centre inside the chest cavity and is protected from impact by the rib cage. It is mainly made up of cardiac muscle tissue which continues to beat from our early developmental stages inside the womb until we die.

Arteries carry blood away from the heart at a high pressure to the tissues, and are generally larger and with thicker muscular walls than the veins which return blood to the heart at a low pressure, with the direction of flow being assisted by the inclusion of valves. Blood includes red and white cells that are suspended in liquid plasma. An adult has about five litres of blood inside their body.

The superlatives of the circulatory system are remarkable:

- Estimates indicate that an adult's arteries, capillaries and veins would stretch 100,000km.
- Animal size affects heart rate (e.g. a blue whale's heart beats around 35 times a minute, and even more slowly when it dives; human hearts beat around 70 times a minute; and the blue-throated hummingbird's heart has been recorded at 1260bpm!).
- Blood can circulate around the entire body in around 20 seconds.
- Red blood cells survive for about 20 days.

- Around three million red blood cells are made and destroyed in the bone marrow every second.
- Arterial blood, which is oxygen rich, is bright red, whereas venous blood, which has depleted oxygen levels, is dark red.

The heart

The heart is a strong muscle that pumps blood around the body. It possesses four chambers as well as valves and blood vessels. The smaller upper chambers (i.e. atria) collect blood and then contract and send that blood through valves into the larger chambers (i.e. ventricles). The heart provides 'double circulation'. It is effectively two pumps that lie side-by-side while working in a coordinated fashion. The heart's right side collects deoxygenated blood from the body and sends it to the lungs. The lungs add oxygen to the blood and remove carbon dioxide from it. The left side collects oxygenated blood from the lungs and sends it around the body. The heart's valves ensure that the blood flows in one direction only. Since the left side does more work its muscle is thicker.

We can listen to the heart rate directly with a stethoscope. or we can measure it indirectly by taking the pulse rate. Our pulse can be felt when an artery is compressed against a bone (e.g. in our wrist, temple or neck). Our heart rate depends on our level of activity, so that when we exercise our heart beats more quickly due to the increased production of carbon dioxide that must be removed, and the oxygen that is needed to maintain the aerobic respiration required in muscular activity.

ACTIVITY

The heart

1. Use a brainstorming technique (e.g. circle time or draw and write) to find out what learners know about the heart and what they think the heart does.
2. Explain to the children that listening is an important observational technique that provides insight into the natural and physical world (e.g. communication and birdsong, erratic mechanical sounds in their bicycle etc.). Listen to normal heart sounds recorded from the Internet. Can they replicate the 'Lub', 'Dub' rhythm by tapping their finger on the table?
3. Use a microphone and speaker to amplify pupils' heartbeat.
4. Use a stethoscope – simple ones work well (e.g. from a doctor/nurse toy set). What can they hear? Try this on different parts of the body.
5. Locate the heart on an outline of the human body.

6   Encourage exploration of a simple pump (e.g. a balloon pump).

7   Use coloured water to demonstrate pumping with a syringe.

8   Compare the pump to the job of the heart in moving liquid forward.

9   Measure the pulse in the wrist, neck or temple.

10 Design an investigation to record differences in heart rate.

The best advice on structuring investigations is to adopt a 'writing frames approach'. These frames place an emphasis on the identification, manipulation and control of variables throughout each investigation, so that predictions, hypotheses, the collection and presentation of data, results and conclusions are managed in a consistent manner and lead to rigorous, fair testing.

An essential precondition for this investigation is that the children will be able to accurately and efficiently take and record pulse rates. Investigations require a lot of planning and so you should consider measuring the heart rate at rest, the number of candidates to be tested, the number of trials carried out by each one, the nature of the activity, as well as the type of exercise and its intensity, the duration of the activity and the rest period, the protocols for taking the pulse rate/breathing rate, and the incorporation of ICT (e.g. a pulse rate monitor, spreadsheets etc.).

Processing results also requires significant levels of planning – how to present raw data, when to use bar charts and when to use line graphs, calculating averages, replicating results, drawing conclusions (do the results match the predictions/hypothesis?), use of scientific language, evaluation of the experimental technique and each stage of the process, reliability.

11 Engage 'carousel brainstorming' to formatively assess attainment of the key ideas relating to circulation. Split the class into groups of four or five members with their own chart and coloured marker. Each group should write down what they know about a sub-topic (e.g. heart rate, pulse, the heart, blood vessels, blood composition, healthy hearts). Place a time limit on the first station and add some time to the subsequent ones that will permit each group sufficient opportunity to read previous groups' comments. Pass the charts around the carousel so that the groups can agree, highlight, question and add to previous recorded answers.

Once each chart has been circulated around the entire class review each one during a plenary, highlighting relationships, evaluating teaching and learning, and planning next steps.

12 Research vaccination.

The human life cycle

Reproductive biology sets out to explore the mechanism by which the species is perpetuated. Sexual reproduction involves the exchange of specialised reproductive cells, which leads to the formation of a new individual who possesses an entirely new combination of traits, inherited from each parent. Genetic variation is the mechanism by which evolution has taken place.

Learning and teaching about human reproduction require that consideration be given to a range of issues that extend beyond anatomical, physiological and genetic considerations; social and emotional well-being, cultural and faith beliefs, and curricular and legislative considerations of impact on these values will prompt clarification during any programme of study. An introduction to sexual reproduction in humans tends to be undertaken during the upper primary stages, however enquiry of all types should be encouraged and supported in an appropriate manner at any stage. Life cycles in various species are covered in Chapter 6, 'Inheritance, Genes and Life'. The main developmental stages in the human life cycle are: foetus (the prenatal stage), baby, child and teenager, adult, elderly person.

ACTIVITY

1   Ask learners to look at photos of themselves to show their progression from newborn baby/toddler/infant, etc. (remember to take care with these and return them afterwards).
2   Make a collection of babies'/children's/adults' clothes and compare these, noting the care and attention given to the design of baby clothing in particular due to their sensitive skin (providing shade and total coverage in full sunlight, and layered, warm, waterproof clothing at other times, since babies' bodies lose heat proportionately faster than children's and adults' due to their larger surface area to volume ratio).
3   A visit to the class – by a baby + mother + grandparent – can be useful in relation to emphasising the developmental stages, and to explore what children can do and babies can't, the levels of care, and to give an opportunity to discuss responsibilities, rights, and personal safety.
4   Compare the recorded voices of people at different ages.
5   Role play to explore the various physical and emotional stages in the human life cycle – as a baby, child and teenager, adult, elderly person.
6   Learners compare the texture, flavour and appearance of baby foods to those that make up their own diet. (Note here weaning, salt, and sugars as issues.)

7 Go to a farm or zoo to see live baby animals.
8 Sequence and label prepared cards that show the different stages of the human life cycle.

Summary

Life processes, vital processes, provide the opportunity to study the way in which living things work – how they obtain energy and process it, how they are mechanically supported in their environment, as well as exploring some of the ways that their movement is brought about. Related learning might be established while learning about sport, dance, and movement. Food and its release of energy and nutrients, and their subsequent delivery around the organism, provides insights into health and well-being as well as animal welfare issues being validated. Reproductive biology can similarly contribute to scientific understanding, especially in the areas of behaviour, genetics and inheritance, while providing an appropriate context for further discussions related to life choices in personal and social education.

REFLECTION POINTS

- Consider appropriate strategies for dealing with 'awkward' questions related to human reproduction.
- Revisit each part of this chapter (Respiration, Obtaining Energy, Support and Movement, the Circulatory System and the Human Life Cycle) and consider the opportunities that could arise for cross-curricular/interdisciplinary/topic work.
- Consider formative assessment techniques that will reveal children's current understanding about what happens inside their bodies. What happens to food? How do I breathe? What is sweat? Where does pee/poo come from? How do I move? How can I stand up? Emotional and psychological questions will inevitably arise within these considerations – how do all these questions join together with other learning priorities (e.g. Health and Well-being, PSE, PE, etc.)?
- Significant confusion exists in distinguishing between respiration and breathing.

  ○ Respiration is a biochemical process which breaks down food to release cellular energy in the form of ATP, and it may take place in the presence of oxygen (aerobic) or in the absence of oxygen (anaerobic).

(Continued)

(Continued)

- ○ Breathing is a physiological process which brings about ventilation of the lungs to permit gas exchange to take place – oxygen dissolves in moist surfaces and enters the blood, while carbon dioxide leaves the blood to be exhaled.
- ○ The source of the confusion possibly lies in continuing use of the terms 'respiratory system' and 'artificial respiration' whereby the latter should be more correctly referred to as 'resuscitation'.
- ○ Which learning and teaching strategies might you adopt to avoid perpetuating this misunderstanding?

- You could perhaps do a quick calculation of how many times your heart beats during the day (or has already done so in your life) – be prepared to be amazed at the totals.

6.
INHERITANCE, GENES & LIFE

Learning Objectives

Having read this chapter you should be able to:

- identify DNA as the chemical basis of inheritance
- review the significance of observation as a scientific skill
- collect source materials and sort these into meaningful groups
- identify inheritance and variation as the product of sexual reproduction
- exemplify discrete and continuous variation, using tallies to record the results
- investigate plant and animal life cycles
- distinguish between inherited and non-inherited characteristics
- debate issues of public concern (e.g. genetic modification)

Overview

It's not that science made history today that is important – scientists make history every day.

Inheritance of characteristics from previous generations is one of the 'big ideas' in contemporary life science. In order to understand inheritance we need to consider a spectrum of related areas: organisms are built from cells; cells are controlled by the nucleus; genes are carried in chromosomes and are coded by the sequence of chemicals in DNA; related species carry identical, related and modified genes.

DNA is inheritance

You might be surprised at the extent of your knowledge about DNA: aspects related to its structure (e.g. genes, chromosomes, nucleus, double helix, phosphate, deoxyribose sugar, organic bases ATGC (adenine, thymine, guanine, cytosine)); historical aspects related to its discovery and the personalities involved (e.g. the Nobel Prize, Francis Crick, James Watson, Maurice Wilkins and 'the dark Lady' Rosalind Franklin); applications in a range of scientific fields (e.g. forensics, medicine and microbiology, genetic fingerprints, the human genome project, genetic modification, therapeutic applications, 'Frankenstein foods'); its significance in contemporary culture and references to DNA both in the media and in fiction.

Significant public understanding exists with regard to DNA and it is no longer regarded as 'belonging' to upper secondary school and university learning. It is commonly referenced material in the media and therefore public awareness is high.

Promoting public understanding and contributing to citizenship are important priorities in school science education.

Reproductive biology involves the production of specialised sex cells, which carry one set of chromosomes from each parent. The evolution of organisms has resulted in particular adaptations to ensure that those sex cells (i.e. gametes) have the capacity to meet each other in sufficient numbers, in the correct conditions and at the right time, to ensure that fertile offspring are provided.

Cell biology allows us to look at the processes that take place during the full life of the cell and the behaviour of the chromosomes during cell and tissue growth. On those occasions when it goes wrong the genes are not expressed correctly, and this results in tissue death or tumour development.

Molecular biology allows us to scrutinise the precise chemical composition of DNA. Francis Crick and James Watson, building on work carried out by Maurice Wilkins and Rosalind Franklin, were able to propose the structure of DNA in Cambridge in 1953. They constructed a model, based on several pieces of evidence, showing that DNA was like a ladder, composed of phosphates, deoxyribose sugar and organic bases. They speculated that the ladder's deoxyribose-phosphate uprights were held together by 'rungs' (pairs of organic bases) – adenine always with thymine, guanine always with cytosine (A=T, G=C) – and the ladder finally twisted into a spiral – the famous double helix.

The identification of 'restriction enzymes' during the late 1960s provided biochemical tools that would permit biochemists to cut DNA into 'fragments'. In other words, they could cut extremely long molecules into short segments and then examine and compare their contents. This includes creating 'genetic profiles'. Profiling also lies at the heart of our contemporary understanding of the genetic relationships that exist between all living and extinct species. Genetic profiles supply compelling evidence that supports Darwin's theory of evolution.

Scientists have generated detailed insights into the chemical basis of inheritance – the publication of the human genome in 2000, and its subsequent refinement, is a well-known example of the extent to which contemporary molecular biology is contributing to pure and applied science. While the Human Genome Project cost the US and UK governments almost $3 billion, the derived insights and developments assured the project's economic success. The costs of genome analysis are now much lower and as of today limited profiles are available for around $150. Scientists aim to provide a complete 'thousand-dollar genome'. The possession of a personal genetic profile has clear advantages in terms of medical care – it does however raise significant issues in terms of ethics and morals.

Molecular biology, and the publication of the genomes of hundreds of organisms each year, are contributing to an intimate understanding of the way those genes work, how they are incorporated into particular organisms, and how identical genes exist in a spectrum of unrelated species. This not only provides insight into the way genes work, it also permits speculation into how these may be modified or adapted, and this is especially important with regard to inherited diseases. This is currently the 'hottest' area in modern life science.

Genetic diversity in turn leads to biodiversity. We will see that two other associated big ideas – *adaptation* and *evolution* – result from the combination of new genes (brought about in turn by sexual reproduction) and the appearance of new ones (the result of mutation in genes and chromosomes.)

ACTIVITY

DNA

1   Given that DNA has such widespread public awareness, and that it is present in all living things, it is appropriate that learners should see how simply it can be extracted from fruit using household resources. This helps to demystify the topic and emphasise its occurrence and widespread availability. Lots of suitable recipes are available on the Internet.

2   DNA modelling can be carried out with a range of resources (e.g. cocktail sticks and confectionery).

3   The family tree provides a worthwhile record to show aspects of inherited and non-inherited characteristics from children's extended family to gather information about an inherited characteristic. Photographic evidence may be used to create a poster. 'Variation within my family', or a family tree drawn in, can show the route of inheritance. Look for the 'Handy Family Tree' at http://teach. genetics.utah.edu (NB Sensitivity will be vital where children have been adopted or are looked after by non-biological parents.)

To see video clips of extracting DNA go to the companion website for the book at **https://study.sagepub.com/chambersandsouter**

Collecting and sorting things

Background

The principal scientific aims of collecting and sorting living things are to be able to recognise differences that exist within a group, determine the differences that exist between them and others, and retrieve that information later.

Observation is a fundamental skill (see Chapter 1 on biodiversity) that provides the gateway to effective learning in science, the humanities, and the arts. The nature of those observations, and the way in which they are cognitively processed, depends on prior experiences, the nature of the object/event under scrutiny, the learning context, and the intended outcomes.

Observation, even at its most trivial level, helps to establish enquiry about the world and ultimately builds understanding. It is clear that children have an innate ability to make detailed observations, and that these can trigger a deluge of interrogation

('What is that?', 'Why does ... ?' etc.). Science education sets out to formalise this skill in such a way that observations and their derived records:

- provide accurate and reliable descriptions
- are precise in aspects relating for example to dimensions and shapes (quantitative – numeracy)
- include detail relating for example to shape, colour etc. (qualitative – literacy)

It is worth noting the high dependency that we have on observations which have been collected by our sense of vision; it is equally noteworthy how strongly language is shaped by our visual observations. Scientists collect data that are unavailable to the senses (e.g. x-rays) and have designed resources to enhance observations and convert them into meaningful data.

The first four lines of Rudyard Kipling's poem 'I Keep Six Honest Serving-Men' can help provide a structure that will assist children to ask meaningful and well-structured questions:

I keep six honest serving-men.

They taught me all I knew.

Their names are 'What' and 'Why' and 'When'

And 'How' and 'Where' and 'Who'.

The skills of observation are important while studying inheritance (e.g. in order to note the variation within, and between, populations). These can be tracked to the differences that exist in DNA between base sequences as the consequence of particular mutations.

Living or non-living?

It is not always easy to determine if something is living and indeed to be convinced by this. Think for example of sponges or coral which have the appearance of being inert, inorganic materials, while they are in fact significant members of marine ecosystems. It is worth considering the seven characteristics of living things – feeding, movement, breathing or respiration, excretion, growth, sensitivity, and reproduction. Children have difficulties suggesting that plants are alive, but will recognise that they can grow. We suggest that several of these areas are beyond the scope of early stages and can be revisited during later years. Determining that something is living by its growth and movement may be sufficient at this stage.

Collections

'The Enlightenment' provided the impetus for the industrial revolution which in turn led to transport, trade and travel around the growing British Empire. Scientific collections were prominent products of these journeys.

While Captain Cook's expedition (1768–1771) aimed to observe astronomical events it was also a discovery voyage, seeking new lands. Joseph Banks, the official botanist, made significant discoveries in Australia and New Zealand. During 40 years as president of the Royal Society he sent collectors around the world. Many cultivated plants are now found in homes and botanical gardens. Exotic animals were collected, displayed and preserved in zoological gardens and museums.

The most significant voyage was Charles Darwin's (1831–1836). His observations, collected all around the globe raised questions that ultimately led to the publication of *On the Origin of Species by Means of Natural Selection*.

The collections of natural history items are vast; over 18 million specimens are held at the Natural History Museum alone. The logistics associated with organisation of such collections are enormous and depend on identification and classification.

ACTIVITY

Collections

1. Brainstorm items that the children collect. What other things do the members of their family collect?
2. Explore how their collections are organised.
3. Research organisational structures within printed catalogues e.g. hardware stores etc.
4. Make a map of the local supermarket. Use colour coding and shading to show the different areas. Visit the site. Rehearse the decisions that you make when you have to locate particular items e.g. a shopping list: bread, apples, sausages, baked beans, biscuits, soap and a shirt.
5. Visit a local museum or library and explore how the collections are presented so that they can be interesting, accessible and entertaining. What is the nature of the collection and does it have a specialist function? What types of resources are available? (e.g. artworks including photographs and sculpture, historical artefacts, geological and biological specimens etc.) What do they have on display? What is on loan to other museums? What is in storage? What resources are available online? How do the public access the collection?

Plant materials

Winter twigs

Collections of winter twigs can provide useful reference materials. These are readily available and can be collected from school grounds or the local environment.

Key features include the colour and shape of the stem and the shape and organisation of the buds. Winter twigs can be compared to identification charts.

Leaves

Leaves provide abundant and usually free resource upon which children's learning about the properties of leaves and variation in particular can be built. Details on leaf preparation and presentation are included in Chapter 1.

ACTIVITY

Leaf collections

1. This activity is best carried out between August and November. Dandelion leaves have a characteristic appearance and when you turn them on the side they have a typical jagged or toothed appearance. Set the children the task of collecting dandelion leaves inside the school grounds, lawns, pitches, beds and pathways during a short period of 5 to 10 minutes. Return to a sheltered area, hall or classroom so that their foraged leaves can be compared. Compare leaves' lengths, widths and the largest 'teeth'.

2. In any collection of leaves, it should be possible to compare their dimensions – lengths, widths etc. comparisons can be made between specimens that have been collected from a single species or between different species. Squared and graph paper is extremely useful here. What is the area of the leaf? Trace around the leaf and use this to calculate its area. This can also be used to measure its perimeter.

3. Organise group research tasks on native broadleaved trees. Potential for differentiation exists as well as giving specific consideration to specific trees in the local neighbourhood.

   Structured enquiry could follow – Common name; Scientific name; Family; Interesting fact; Overall shape; Leaves; Flowers; Fruits; Distribution; Associated animals; Mythology; Commercial uses; Height estimate (see Chapter 2); Circumference at 1 m.

4. Conventions in numeracy determine standard units of measurement.

   Children might be aware of the ideas associated with arbitrary establishment of those conventions on the basis of e.g. the length of the King's arm. Invite the children to find out 'how tall you are in leaves' by using a leaf from 'their' tree as the standard unit of measurement. They could make a measuring stick e.g. equivalent to the length of 10 leaves; they may wish to use another base number. They can lie flat on the ground and use a leaf and the leaf measuring stick and count how many of them makes their partner's height.

   Invite discussion on accuracy, units of measurement, base numbers and so on.

Wild and garden flowers

Garden flowers are essentially cultivated varieties of the plants that have been collected from all over the world for their attractive properties. They have been selectively bred to vary size, colour, scent and any other desired feature. Seed catalogues boast their special qualities.

Keys

Taxonomy involves the systematic organisation of living things into categories that demonstrate their relationships to each other. Additionally the grouping implies evolutionary proximity and trends. Taxonomy is an essential if somewhat neglected area of modern academic biology, yet areas associated with today's 'big issues' of biodiversity and the contents of the gene pool depend on an accurate identification of living things. The skills required are significant yet the numbers of specialists are falling, with expertise frequently located in enthusiasts who comprise particular interest groups.

Simple biological keys might be used to identify broad groups (e.g. flowering and non-flowering plants, vertebrates and invertebrates, fungi or other organisms). A biological key – a dichotomous key – is a tool used to determine the identity of an organism by answering a series of questions about that organism's characteristics. Dichotomous keys usually include a series of questions, each of which has two possible answers. By correctly answering these questions in sequence, the responses lead to an end-point that determines the identity of the organism.

Simple keys that are provided for non-specialist use have notable shortcomings. Of necessity they provide brief descriptions and illustrations of a restricted list of possibilities. Their incomplete data therefore risk confusing teachers and pupils rather than providing the intended clarification. Variation in populations and between species, as well as the potentially overwhelming nature of biodiversity, place teachers and learners in almost impossible situations. While a child in the UK may be able to recognise a plant in the school grounds as being a moss, to which of the more than 750 identified species does it belong? Similar frustration can arise with wildflowers (e.g. Francis Rose's *The Wildflower Key* for Britain and Ireland extends to almost 600 pages).

Despite these reservations, the logistical exercises associated with making and using keys do provide useful learning experiences in relation to data management.

Well-developed skills of observation and description are necessary for using and making keys. Several different types of keys are commonly used to identify the entire range of living things – branched, paired statement (dichotomous), and lateral keys. While it is important that learners should be able to use a key accurately, by making careful observations, it is essential that they should also be able to directly view living things in the classroom and its immediate surroundings. A growing sense of wonder should exist at the variety of living things that surround us – teachers should not feel threatened if they are unable to answer pupils' questions in the field on the identity of living things, it is instead more desirable that they should celebrate the recognition of something new (to them at least) and use the resources to hand to make a tentative identification, or at least show it to another pupil so that they may be able to find the same thing. It is essential that the widest range of observations is made: encourage learners to LOOK, SMELL, and FEEL living material in order to extend their observations and promote their descriptions. (Remember though that taste is inappropriate for obvious safety reasons, and that handwashing is an essential aspect of good practice in science.)

Pupils' observations and descriptions require explanation – this can promote not only language development, contributing to literacy, but also scientific excellence. Pupils also need to be encouraged to make explicit their reasons for grouping/classifying as they will not do this automatically. Fresh and living material will provide first-hand experience and should be used as much as possible. Drawings and photographs are useful secondary resources but cannot show all the significant details. They can only supply a visual record (which is frequently a poor substitute).

Using keys

Pupils' abilities to use keys will depend on the accuracy of their observations, the quality of the key, and its appropriateness to their stage. The discipline of taxonomy relies on the availability of accurate keys. Many of the specialised keys are held on the Internet. *The Tree of Life* is a collaborative web-based project, one that is maintained by biologists from around the world. It is intended to provide insight to an international, public audience and includes information that is of interest to everyone.

Pairs of differences between leaves

Individual leaves can also be classified as either simple or compound (i.e. several or many leaflets per leaf).

Figure 6.1 *Simple leaf*

Figure 6.2 *Compound leaf*

Simple leaves have a single leaf blade. Compound leaves have several leaflets attached to a central stalk or midrib. This stalk is not woody and no buds occur at the base of the leaflets.

Leaf margins (i.e. the edge around the leaf) can have many different forms (e.g. serrated or smooth). The shape of the leaf is very important in helping to identify a particular tree. Leaves in the same family will, sometimes, look very similar and have similar common names.

Buds grow on either opposite or alternate sides of a twig. This is very helpful during winter tree identification.

Classifying animals and plants

In addition, pupils can move on to researching the ways in which plants and animals are classified, working through the various levels from kingdom to species – though not necessarily including all of the levels.

Collections of flowers, pond life and soil organisms supply opportunities for identification activities. Note that the collection of living things must conform to good conservation and safety practices.

Making keys

Key construction depends on the combined skills of making careful observations, organisation and subsequent recording. The process starts by identifying the similarities and differences that exist between collections of objects:

- Keys draw on the differences between the physical characteristics of animals and plants.
- When you are making a key, sort the objects into groups according to their main characteristics (e.g. shape, size, colour). (Physical sorting can help.)
- The differences need to be put into order of importance and used as decision points in the key.

We suggest that key construction should follow using keys once learners have established their familiarity with the various kinds of keys used in identifying living things. This helps in establishing the need to look carefully for similarities and differences in their physical characteristics.

Note that branching keys are more straightforward to construct and that paired statement keys should only be attempted once pupils have mastered constructing branching keys. Conventions must be applied in paired statement keys, and these include numbering, choices and pathways:

- Number each statement sequentially.
- Continue dividing into smaller groups.
- Either name each individual or direct towards a subsequent question.

ACTIVITY

Keys

1 Make a key to identify invertebrate collections from soil or pond animals (photographic records can help here).
2 Construct a key to identify routes inside the school building.
3 Construct a key to identify all class members (avoid any features that could be discriminatory or lead to bullying).
4 Pupils may invent their own creatures and create keys to identify individual members, highlighting significant differences in physical characteristics.

Inheritance – it's all to do with sex!

Our understanding of inheritance is based on a few key ideas:

- Genetic information is stored in the nucleus of cells as chromosomes which are made from DNA.
- Half the genetic information comes from each parent.

- The appearance of living things is affected by inherited characteristics for genes and environmental factors such as light, diet, and amount of exercise.
- Inheritance leads to variation – which in turn leads to adaptation and this drives evolution.

Genetic information is passed on from a parental generation to its offspring. This information is located in the DNA of the chromosomes of the parental sex cells (gametes). Each chromosome is made up of a number of genes. Each gene controls a specific characteristic and this might lead to an aspect of the offspring structure (appearance) or its biochemistry (function) through the production of specific proteins. Each parent provides one set of genetic information in the form of chromosomes. These fuse with a complementary set from the other parent at the point of fertilisation. Thus information from each parent combines to produce a new organism with characteristics inherited from each parent. Similarities and differences arise in family lines – some offspring are strikingly similar to their parents and others are notably different.

Sexual reproduction results in new combinations of genes and geneticists refer to this as recombination. Principles and laws allow them to predict the nature of offspring and the possibilities. This lies at the basis of the selective breeding of plants and animals for livestock or crops, and also permits prediction by genetic counsellors.

Work in this area is governed by probability and explicit links can be made to learning in mathematics.

Body cells (somatic) contain matched pairs of chromosomes. Different organisms have different numbers of chromosomes: horses have 66 chromosomes (33 pairs), humans have 46 chromosomes (23 pairs), onions have 8 chromosomes (4 pairs). Note that there is a species of ant which has only a single pair of chromosomes, while the record for the highest number has held by a tiny fern which has been found to have 630 pairs of chromosomes.

One pair of chromosomes in humans controls the inheritance of gender and this is called sex determination. In females the sex chromosomes are identical and so these are labelled XX. In males the sex chromosomes are different and therefore these are labelled XY.

ACTIVITY

Genetics

The predictive nature of genetics can be illustrated with the following activity – a role play that sets out to provide opportunities to calculate simple probability. These can be expressed as fractions or in words.

Prepare or purchase some party hats – four pink ones to represent the female and four blue ones to represent the male. (This is intended to provide 'flagging' rather than promoting gender stereotypes.) Each hat represents a set of chromosomes, each pair represents a parent. Inside each pink hat hide an 'X' and inside two blue hats place an 'X', and the other two place a 'Y'.

The role play involves simulating the production of sex cells, the random nature of fertilisation, and illustrating the offspring (see Figure 6.3).

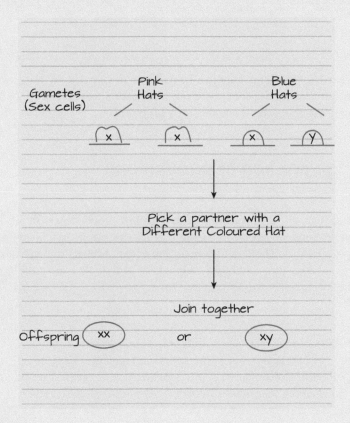

Figure 6.3

(Note 1:1 or 50:50 or percentage offspring and that these proportions can be predicted.)

(Continued)

(Continued)

Repeat the role play with four hats to illustrate the increased numbers of combinations and that the offspring proportions remain unchanged. Note that several other genes will be located on these sex chromosomes that will be inherited alongside gender. Invite speculation about the huge numbers of combinations that are possible with 23 pairs of chromosomes in the human genome. (Recognising of course that each chromosome contains hundreds and possibly thousands of genes.)

Reebops

Model organisms for teaching genetics have been developed by Patti Soderberg at the University of Wisconsin. These usefully illustrate several aspects of inheritance including predicting the genetic make-up of offspring.

Describing variation in populations

Variation

If we look at any group, a population of living things including people, we will see many differences between them. For example, all the pupils in your class will look different from each other. They will however have lots of things that are the same (two arms, two legs, a nose and a mouth) and lots of things that are different too (hair colour, eye colour, height, and differently-shaped noses).

Five ideas

- Variation – the observable differences between individuals.
- Inheritance – when parents pass on genetic information to their offspring.
- Genetics – inherited characteristics are genetic since genes pass on the information.
- Genetic information (genotype) – coded in DNA that is stored in the nuclei of cells within the chromosomes.
- Phenotype – appearance of living things is affected by inherited characteristics from our genes and environmental factors.

Each gene controls a single characteristic (e.g. Rhesus blood group) but others (e.g. hair colour and eye colour) are controlled by several genes. Each parent provides one set of genetic instructions in the form of chromosomes. These fuse with a complementary set from the other parent at the point of fertilisation. Thus information from each parent combines to produce a new organism with characteristics inherited from each parent. Similarities and differences arise in family lines – some traits can be observed through several generations of a family line.

Environmental factors such as health and nutrition are significant in determining the ways that genes are expressed. Each can change the appearance of the organism. Geneticists talk about genotype (genetic makeup) and phenotype (appearance as a result of the impact of the environment on the way that the genes have been expressed.)

ACTIVITY

Differences

We suggest a variety of activities to introduce the existence of differences (variation) between individuals of the same species. Take this opportunity to stress the fact that we inherit genes from our parents as they did from theirs.

1   Plants can be used to illustrate the effect of the environment on individuals. Germinate seeds with or without light. The seedlings without light will be yellow (etiolated) and will grow tall and spindly.
2   Similarly you can grow seedlings with a sufficient water supply and in drought conditions. After a week or two, there should be clear differences in the sizes of the plants. As an extension, it would be interesting to water the droughted plants to see how long it takes for these to recover, if at all, from the lack of water.

Types of variation

Different types of variables exist in nature that can be described in qualitative or quantitative ways. Continuous variation occurs when small, measurable differences exist between individuals. Discontinuous variation is a result of discrete, qualitative differences between individuals. Teachers should note that the type of variable determines the nature of the record. The potential to link variation strongly to numeracy is self-evident.

Continuous variation

Continuous variation is found where there is a gradual change in a characteristic between individuals. Common examples include height or age. Continuous variation is mathematically defined and will show a normal (or Gaussian) distribution in a population. Values range between extremes, but the majority of the population will be at the middle of the distribution – this gives the characteristic bell-shaped curve when a frequency graph is produced.

Discrete/discontinuous variation

'Discrete' or 'discontinuous' variation describes the characteristics that show distinctive differences. Variation is described as 'discontinuous' when a population shows few or no intermediate phenotypes (appearance) between the extremes. These are frequently either one type or another (e.g. freckles/no freckles), but in some cases (e.g. blood types) intermediate forms are found. Pupils normally find it fairly easy to demonstrate examples of this type of variation. Other examples of discontinuous variation in humans include tongue rolling, type of earlobe, and eye colour.

ACTIVITY

Discontinuous and continuous variation

1   Observation and collection of data from the class/year will reinforce the extent of discontinuous and continuous variation which is apparent. Start with these.
2   Celebrate diversity and be prepared to be entertained by the unusual things that might appear (e.g. clover leaf tongue).

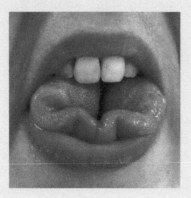

Photo 6.1

© N Souter

Differences in continuous variation are normally easily measured (e.g. shoe size, height, weight, circumference of wrist, hand span, length of index finger). Bear in mind the nature of the data – qualitative or quantitative. (Learners can score themselves on half a dozen simple genetic traits, most of which are determined by single genes.) Maintain sensitivity towards the inclusion agenda (with regard to height, weight, etc.). Multiple gene traits such as eye and hair colour are difficult to relate to and are best avoided.

3　Arms and hands – if you ask the children to clasp their hands, you'll find that some will clasp them with the left thumb on top and others with the right thumb on top. Ask the children to clasp their hands or fold their arms 'the other way round'. How do they feel when they do this? Record the results and then find out how many fold their arms with the right arm on top compared to those who fold their arms with the left one on top. Do any relationships exist between arm folding and hand clasping?

Photo 6.2

© N Souter

Photo 6.3

© N Souter

4　Data can be collected within school populations and recorded over several years to generate trends and patterns.

5　Investigate the variation that exists in 'spreading' fingers or spreading toes. The angle can be measured, and combined with mathematical work as shown in the photo.

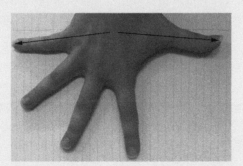

Photo 6.4

© N Souter

(Continued)

(Continued)

6   Plant material is especially useful here. Measure, for example, sunflower seed length, privet leaf length, or other continuous variables; look at discrete variation in for example flower colour in parks and gardens (seed and bulb catalogues can provide useful secondhand resources).

Photo 6.5

© N Souter

These activities also give an opportunity to introduce making a tally chart to record the data, and following the convention of making a stroke for each count and making the fifth one a diagonal across the first four to give a 'farm gate' appearance.

7   Variation within banded snails has been well studied, and so the potential exists for children to engage in 'Citizen science' projects (e.g. for an evolution mega lab, or to establish a data set within the school for future study).

Life cycles

The life cycles of living things provide interesting and varied learning experiences for children, especially if these involve first-hand, direct experiences.

Their own growth is often a source of great amusement but can also provide useful data. Photographic records also provide insight into children's growth and development during an extended period – wardrobes have the potential to supply similar developmental reminders.

A notable feature of mammalian life cycles relates to the extent of parental care until offspring reach maturity. Animal behaviour is frequently determined by the reproductive cycle and might involve migration, courtship rituals and driving previous offspring away, as well as the extent of maternal and paternal involvement.

The life cycles of other vertebrates, including incubation, eggs, internal and external fertilisation and parental care, help to configure teaching programmes. Two main styles occur in the life cycles of invertebrates – complete metamorphosis (e.g. butterflies) and incomplete metamorphosis (e.g. stick insects).

Classroom collections can supply engaging resources that will provide long-term opportunities for classroom studies. Stick insects and brine shrimp are straightforward to maintain in the classroom and illustrate good examples of incomplete metamorphosis. Stick insects require fairly minimal attention, can be maintained, cleaned and fed by willing classroom volunteers, and demonstrate the life cycle – starting with eggs, progressing through several stages, and leading to the egg-laying adult stage. An additional extraordinary feature of stick insects is that adult females will lay eggs without being fertilised. Butterflies, bees, flies and mealworms are all examples of insects that involve complete metamorphosis. The egg hatches into a grub, caterpillar or maggot which enters a resting phase (chrysalis), and then emerges as a mature, reproductive adult. The most significant thing about these life cycles is that the adult and juvenile stages bear a minimal resemblance to each other.

Artemia (brine shrimp) or sea monkeys can similarly illustrate several different stages, where the larvae simply appear to become larger over several steps until mating adults can be seen swimming, coupled in the saltwater. (Further aspects relating to life cycles in plants are outlined in Chapter 4.)

ACTIVITY

Collection study

1   Identify opportunities and establish collections of living things that will permit the ethical establishment, maintenance and upkeep of populations of invertebrates in the classroom.
2   Visit a butterfly farm.

Areas of public concern

Science and religion

Children's ideas about 'Life, the Universe and Everything' are under constant review.

Their views are influenced by the widest range of factors that can lead to confusion and cognitive conflict. This results in questions that teachers can find challenging. They recognise that their family, teachers, friends and the media each contribute to children's learning. Children's faith in particular sources during the early years can also be unquestioning (e.g. 'the teacher says so' can settle many playground disputes). Learning science involves developing a series of skills that involve observations, hypothesis testing, experimentation, etc. Faith and science are alternative philosophies.

Science and religion provide challenging ideas, notably where host communities' views oppose scientific ideas, in particular those relating to Darwin's theory of evolution (e.g. the creationist movement). The associated undermining of the science curriculum to provide a 'balanced' coverage between creation and evolution is an example of curricular, not scientific controversy.

Such conflict need not exist. It does not exist in science, and nor does it exist in most religious thinking.

The *scientific evidence* to support evolutionary theory is overwhelming; *belief* of the existence of God is accepted by all major world faiths. Each describes different philosophical viewpoints. While science changes our perceptions of the natural world, which is perhaps its prime purpose, it cannot be used to prove or disprove the existence of God through experimentation. On the other hand, the rhetoric of religious questions lies in different philosophical domains, and frequently invites questions that seek to explain our existence.

Genetic modification

Anything that leads to a new sequence of genes can be considered as providing genetic modification. It is clear therefore that sexual reproduction, and the associated independent assortment of chromosomes and genes, provide genetic modification – we are as different and similar to our parents as our children are from us.

Civilisation has been brought about by our human capacity to manage the genetic modification of plants and livestock. Selective breeding (or an accident in nature) led to the appearance of the first varieties of wheat – these resembled our modern 'Durum' wheat. The first ancient civilisation grew around the ability to harvest crops and be released from the hunter–gatherer existence, and this gave rise to the establishment of the first villages and towns. Selective breeding for favourable genetic traits has been the farmer's skill throughout civilisation. Recent advances in biotechnology now permit individual genes to be transferred between varieties and between different species – and therein lies the controversy. Viewpoints range from this being an exciting and acceptable scientific advance that will serve mankind, to accusations of 'playing God'.

Scientific knowledge is advanced through a critical examination of the results of investigations by other scientists. Scientific facts, hypotheses and theories are regarded as being tentative, debated and revised. Public controversy involves various agendas.

Children could be encouraged to further their understanding by investigating or debating any relevant and interesting topic. Examples could include gene therapy, pharming, transgenic animals and plants.

Controversy in society

As people we will not always agree with one other: we might prefer different food from our friends; support a different football team; enjoy different music or art; have different viewpoints on political issues. Being 'controversial' involves much more than simply disagreeing or expressing a matter of opinion.

Scientists work in particular ways to test their *theories* in order to confirm their *hypotheses*. This involves seeking *evidence* to support the argument and publicly debating it over a long period. When they *interpret* the results a number of different *conclusions* might be reached. Scientists also publish their research so that other scientists might challenge and discuss every aspect of the process. This is called peer review.

Peer review and debate around published research continue to challenge scientific understanding. Controversy in science happens when scientists disagree about the suitability of conclusions. It is an essential part of the way that science progresses through the interpretation of evidence, and the discussion and debate about meaning.

The nature of science

> Science is a way to teach how something gets to be known, what is not known, to what extent things are known (for nothing is known absolutely), how to handle doubt and uncertainty, what the rules of evidence are, how to think about things so that judgements can be made, how to distinguish truth from fraud and from show. (Richard Feynman, Nobel Laureate)

Scientific knowledge is supported by evidence. However, when new techniques or processes become available and ideas are constantly tested, then new knowledge will appear and replace the existing ideas. For this reason *all* scientific knowledge is regarded as being *tentative*.

Scientific knowledge is built upon facts, hypotheses and theories. Scientists regard *facts* as something which can be observed (e.g. the length of this book). A *hypothesis* is a statement that takes into account previous observations and can then be scientifically tested. *Theories* in science are broad ideas that help to make sense of related observations and experiments.

Scientific methods involve lots of different approaches. Scientists work individually or within teams; they test hypotheses; generate different forms of data; repeat their experiments; reach conclusions; and take part in the peer review process. There really is no single way of working or scientific method.

Controversy in science

It should be clear that an essential part of being scientific involves constantly challenging new scientific findings and established 'facts'. Scientific journals are full of correspondence where controversy appears when scientists debate different viewpoints.

You would think with all the publicity in the press and social media that topics such as evolution by natural selection, animal testing, nuclear power and climate change would be controversial among scientists, but that is not the case.

Evidence to support Charles Darwin's ideas about evolution by natural selection is drawn from all areas of biology and contributes to their progress. The science is accepted; it is not controversial.

Scientists do not rate animal testing as being controversial. One study reported that '... 93% of scientists were in favour of animal research, merely 52% of the public agrees'. Once again, the science is accepted; it is not controversial. Animal testing is carefully regulated by the government. More than 500 distinguished UK scientists, including three Nobel prize-winners and 250 university professors, pledged their support for animal testing in medical research.

> Throughout the world people enjoy a better quality of life because of advances made possible through medical research and the development of new medicines and other treatments. A small but vital part of that work involves the use of animals. (Research Defence Society, 2005)

Scientists are also more in favour of nuclear power than the general public, and while nuclear accidents have caused shocking results, it should be noted that many more people die from the effects of air pollution each year. Again, the science is accepted; it is not controversial.

Climate change is another area where scientists accept that this has been brought about by human behaviour – the science is accepted; it is not controversial.

It must be true – it was in the news!

Keep your wits about you on 1 April! Newspapers frequently provide stories that appear to be true, but they are having a laugh and simply trying to fool you. Other news might have more cynical purposes, and skills are required to consider the authenticity of the assertions contained within reports. But how can we evaluate the accuracy of a report?

ACTIVITY

Evaluating accuracy

1　Encourage children to scrutinise media reports by being critical and applying scepticism, doubting, seeking evidence, and undertaking further/alternative enquiry.

2   Encourage children to be argumentative in supporting their presentations – or at least be argumentative in scientific terms, and to recognise that scientific arguments are not about having a row – they involve supplying reasons where claims are justified on the basis of data.

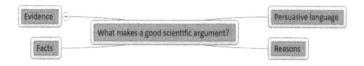

Figure 6.4

Summary

The significance of DNA is such that a Google search of 'DNA' yields about 315 million results. DNA appears in the news on a daily basis. It is worth recollecting that it was not until 1953 that its structure was correctly described, and that while it led to the scientists Crick, Watson and Wilkins being awarded a Nobel prize, it has been its subsequent management, modification and manipulation that have made it the most culturally significant chemical in history. Applications include gene profiling, which has forensic and diagnostic uses; in cloning and genetic insertion, which are important tools for breeding and therapy; and for shedding light on further scientific insights into the mechanisms of inheritance. We hope learners will be able to enter discussions on aspects related to controversy in the application of DNA technology. While acknowledging the importance of DNA it is also worthwhile noting that secure learning in this area is derived from children's capacity to make accurate observations, record these, and sort them into meaningful groups. Exciting opportunities exist for demonstrating heritable traits and in using statistical discussions to distinguish between discrete and continuous variation.

REFLECTION POINTS

- Consider the scope that exists to scrutinise topical science issues that arise within your class and within your programme of study.
- Write 'DNA' in the centre of a sheet of A4. Set a timer for two minutes. List everything you can recall about DNA.

(Continued)

(Continued)

- What resources exist locally where you could collect suitable materials for children's study? How might these be preserved and stored for future use?
- Explore the availability of online resources that would support the ideas associated with the genetic changes that result from sexual reproduction.
- Identify suitable materials for lessons on continuous variation; link these to numeracy targets.

7.
ANIMAL BEHAVIOUR

Learning Objectives

Having read this chapter you should be able to:

- describe some features of the human sense organs
- explain some facts about the way they work
- describe some of the limitations of the human senses
- extend children's observational skills by using the full range of senses
- research examples of innate and learned behaviour
- look at behavioural adaptations (e.g. swarming, migration and communication)
- carry out practical investigations
- carry out investigations with live materials (e.g. habituation in snails)
- provide examples of group, cultural, social and territorial behaviour

Overview

The behaviour of animals has been studied since ancient times as it helped early man to hunt suitable food and collect animal products.

Behavioural adaptations help animals to survive in their environment. Innate, learned and social behaviour depend on effective communication that takes place through a variety of different approaches, including sound, chemical and visual signals.

Children will probably be familiar with several examples of animal behaviour from watching local wildlife, family pets, or natural history broadcasts. Learners can be encouraged to explore animal behaviour by making careful observations and testing hypotheses related to:

- researching examples of innate and learned behaviour
- looking at behavioural adaptations
- carrying out practical investigations
- carrying out investigations with live materials
- providing examples of group, cultural, social and territorial behaviour

Living things in the classroom

The justification for keeping living things in classrooms extends beyond providing a stimulating learning environment; they are much more than 'wallpaper'. Children have a great interest and enthusiasm for looking at animals. Living things are wonderful resources that provide children with opportunities to nurture, to see, feel and touch animals in responsible, caring and respectful ways that provide connections with the wider world. Close observations are possible in an environment that is controlled and safe for the living things as well as the children. Various studies by the Waltham Centre for Pet Nutrition have demonstrated significant affective gains through children's engagement with animals, notably in the areas of self-esteem, empathy, social engagement and social cohesion in particular.

Animal welfare

Animal welfare issues relating to environment, diet, opportunities to exhibit normal behaviour patterns, and the avoidance of discomfort, distress or disease must be taken into account during planning. Principles related to 'duty of care' suggest that maintaining vertebrates is unfeasible in a school environment where they may be unattended over weekends and holidays. The RSPCA strongly discourages keeping such animals in schools. Legislation and local guidelines need not create obstacles to achieving the affective gain mentioned previously.

Visits to farms, zoos, wildlife parks and other animal collections, where the capacity exists to ensure the highest standards of animal welfare, can provide enriching learning experiences. Bird feeders also provide ethically appropriate opportunities for children to observe animal behaviour directly from the security of the classroom, and where animal welfare is a priority.

While vertebrates would be unsuitable in our classrooms, the scope for maintaining populations of invertebrates (e.g. stick insects, snails, pond organisms, seed beetles, brine shrimps), and a variety of plant materials, should be encouraged, having the potential to enrich the learning environment. It is essential that the invertebrates are handled in a suitably caring manner that avoids the integrity of their outer body (exoskeleton) being compromised in any way, since this would lead to a loss of body fluids and their death. Careful handling is vital while moving them for close examination.

Types of behaviour

There are lots of different types of animal behaviour. Some behaviour – innate behaviour – is inherited while other behaviour is learned. Certain behaviour is shown by individuals, such as a frightened animal running away or a dog panting, while other behaviour involves cooperation between individuals or social behaviour. Examples of social behaviour include honeybees working together in a hive, animals mating, carnivorous animals hunting in packs, and parents sharing care of their offspring.

All animal behaviour is designed to help animals respond to their environment in order to survive. Animals respond to their internal or external environment. Internal cues include hormones, nutrients and temperature, while external ones include day length, temperature and time. Animals use their sensory system to detect changes in their internal and external environment. It is the brain, however, that determines the behavioural response.

Innate behaviour

Automatic, inherited behaviour is based on instinct and no conscious thinking is involved.

The male European robin (*Erithacus rubecula*) defends his territory instinctively by attacking potential competitors. During the mating season he will also attack toy birds or even a few red feathers. The red colour is the trigger for this aggression. The survival value is that the cock robin will be able to breed and thus pass on his genes.

Baby turtles instinctively make their way towards the ocean. Here the survival value is related to avoiding predators and obtaining food.

Humans also show innate behaviours but these are mainly restricted to babies who will suckle on a nipple instinctively. This helps their survival by providing food. Another innate behaviour in newborn babies is the grasp reflex. A baby will instinctively grip anything that is placed in the palm of their hand. This grip can be surprisingly strong. It seems a plausible explanation that this reflex is the evolutionary remnant of the grasping reflex seen in other primates, that is used to ensure the infant primate clings to the mother's hair while being transported from place to place.

Learned behaviour

Our pets soon learn where their food comes from. They use their senses to detect where it is stored and use their memory to shape their behaviour. Learned behaviour has the advantage over innate behaviour because it is dynamic and can respond to changing conditions. Most cats will know where neighbours keep food and treats!

Although all animals are able to learn, humans are the most intelligent animals, capable of learning more than any other species. Indeed lots more! Nearly all human behaviour has been learned. Learned behaviour is the result of previous experiences. Humans learn all the time. Learning is much more than what children do in school. Survival depends on learning and this often means at the most basic level – on satisfying human needs such as food, warmth and shelter. Once these basic needs have been fulfilled, humans are free to learn any number of wonderful things, from riding a bike to space travel, programming a computer or playing a musical instrument etc.

Adapting behaviour

Social behaviour

Animals interact with other members of their own species and exhibit social behaviours.

Honey bees (*Apis mellifera*) provide one of the best studied examples of animals exhibiting social and cooperative behaviour in nature. They live in large colonies with as many as 50,000 workers supporting a single queen. The worker females undertake tasks progressively throughout their lives, and cooperate to make wax cells, feed, clean and nurse the larvae, control the hive temperature, make honey, defend the entrance from predators, and collect pollen and nectar to feed the entire colony. The queen lays eggs while the male drones only exist to fertilise her.

Most honey bee behaviour is innate, however the workers are able to learn where to find their food and they can communicate this information to each other. In favourable conditions new queens are produced and the colony will 'swarm'. Although a swarm looks quite frightening, the bees themselves are in fact full of honey, looking for a new home, and quite calm.

Social behaviour however is not always cooperative. During the rutting season male red deer will compete with one another in dramatic head-to-head battles over the right to mate with herds of females.

Migration involves the seasonal movement of populations of animals, usually in response to the need to find food. Examples are found throughout the animal kingdom – mammals, birds, fish, and many invertebrates including insects, squid and starfish.

Barnacle geese have been extensively studied by scientists. They spend the summer breeding season inside the Arctic Circle and take advantage of the long day length to provide food for their chicks. The entire population migrates south in the autumn to avoid the cold Arctic winter. Up to 30,000 Barnacle geese migrate between the Arctic Archipelago of Svalbard and the Caerlaverock Wetland Centre, Solway Firth, Scotland – a round journey of more than 6000km – each year.

The geese travel in large 'V'-shaped formations, called skeins, which not only allows the migration to be coordinated but also increases the flock's overall flying efficiency. The geese are able to fly further and for longer periods due to this formation.

ACTIVITY

Social behaviour

1. Visit an apiary and observe a demonstration of social behaviour in the hive.
2. Record populations of birds feeding at the birdtable at different times of the year.
3. Research migration in selected mammals, birds, fish, reptiles, amphibians, and insects.

Communication

Sound

We are all familiar with the sounds that are made by our domestic pets as well as the birds that inhabit our gardens. Animals use communication to send messages and warnings about many things, including mating, food or predators. Acoustic information depends on the animals being able to make and detect particular sounds.

Sounds are also used to assist navigation in whales, dolphins and bats, but the sounds they use are often outside of our hearing range. Bats are able to navigate successfully in the dark and trap their prey by using an ultrasonic signal.

Low-frequency vibrations help honey bees to pass on information about their exact location – the direction, distance and abundance of food – using a 'waggle' dance that is detected by other foraging bees in the hive.

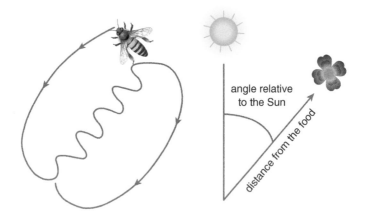

angle relative
to the Sun

distance from the food

Figure 7.1

Vision

Visual signals are used by lots of animals to communicate with each other (e.g. mating displays in birds). Dog owners will be familiar with a range of postures that communicate their pet's emotional status (e.g. when it is yawning, blinking and nose licking, compared to when it is standing with its ears back and tail tucked underneath). A number of animals use facial expression as a form of communication. Chimpanzees are a very good example of this.

Other animals use colours in different ways to attract mates or warn predators. Male peacocks display their impressive colouration to attract females, while female glow-worms communicate via light produced in their tails to attract a mate.

Chemicals

Molecules that are used for chemical communication are called pheromones. These chemicals can be used to identify a mate, mark territory, or spread alarm. Pheromones are well studied in insects in particular. Ants for example will leave pheromone trails to assist the colony in locating rich food supplies. Other pheromones provide a chemical 'key' that permits access to a beehive; insects that attempt to enter the hive and lack this precise chemical are attacked by sentry bees. Pheromones are especially important in attracting mates and these can be used to trap pests (e.g. clothes moths). Otters communicate with each other and mark their territory by 'sprainting'. This provides other otters with a chemical clue of their presence. We can also infer from the success of the perfume industry that pheromones play a significant part in human behaviour.

ACTIVITY

Behaviour

Watch *Let's Ask The Animals*, a short, entertaining movie produced by the Association for the Study of Animal Behaviour that is widely available on the Internet and ' ... looks at farm animals and the processes they share with humans. Comparing basic needs, including companionship, nutrition, exercise, keeping clean, it finds many similarities'.

Habituation

An animal may respond defensively to something when it first encounters it, but over time their response to that stimulus will reduce. The gradual fading of a response to stimulus in this way is known as habituation.

Scarecrows have traditionally been used by farmers to frighten birds away from their crops. As a result of habituation, however, the birds become used to the sight of the scarecrow and so are no longer scared by it. The most effective scarecrows are lifelike and will move and make sounds. Unfortunately, however, habituation by birds is inevitable, so the 'power' of even the best-designed scarecrow will diminish over time.

ACTIVITY

Habituation in snails

You may have noticed that when you touch a snail it withdraws its eye stalks.

(Continued)

(Continued)

Photo 7.1

© N Souter

(Note that the gentlest touch is required here and so the opportunity arises to discuss our ethical treatment of animals and responsible practice during experimentation. Note also that we must not touch the eye stalks, only the flesh at the front of the snail's head.)

The aim of the experiment is to find out if the snails will become habituated to being touched. How would you carry out this investigation? What would you expect to see if the snail became habituated to touch?

The Owl That Was Afraid of the Dark

Jill Tomlinson's 1968 book is a classic piece of children's literature. The irony contained within its title relates to animal behaviour. This book has generated abundant examples of good practice in topic studies work across a wide range of curricular areas. It has even been recommended to assist clinicians' treatment of children who have a fear of the dark.

ACTIVITY

Undertaking research

1   Research owls. Where do they live? What is their food? What 'superpowers' do they possess?
2   Research how sounds can be directed towards the ears by external anatomical features (e.g. the disc shape which is characteristic of owls' faces).

3 Owls swallow their prey and regurgitate indigestible materials (e.g. bones, fur and feathers) in a pellet. Dissect an owl's pellet to find the remains of its prey.
4 Make a list of nocturnal and diurnal animals.
5 Investigate adaptations in owls' eyes for night and for binocular vision.
6 Compare and contrast different owl species (e.g. Tawny Owl, Short-Eared Owl, Long-Eared Owl, Little Owl, Barn Owl). Structured reports could make reference to shape, size, habitat, diet, population, scientific name etc.
7 Visit a planetarium or observatory; invite local enthusiasts to bring telescopes to the school during clear winter evenings. Observe constellations.

Earthworms

Earthworms are easily collected from gardens, compost heaps or school grounds, and can also be purchased from angling stores.

Their subterranean lifestyle means that a lot of their behaviour is poorly understood. Anatomically the earthworm is a segmented tube where each section surrounds the alimentary canal. Its 'brain' and 'heart' are found in the anterior (front) of the animal. Although they have no eyes they are sensitive to light. They lack ears but are sensitive to vibrations in the ground.

ACTIVITY

Earthworm observation

1 Observe an earthworm moving across a ceramic tile. Look for peristaltic movement.
2 Listen carefully as the earthworm moves inside a rolled piece of paper.
3 Bring a cotton bud close, just touching the anterior and posterior ends of the earthworm.
4 Soak a cotton bud in vinegar and bring it close to (i.e. not touching) the anterior and posterior ends of the earthworm.
5 Design an experiment with simple resources to test the earthworm's sensitivity to light.
6 Place a small mass of soil in one half of a dish and vermiculite at the other – does this influence earthworms' movement?
7 Construct a wormery from simple resources and inoculate it with earthworms. Observe any changes in the wormery during the weeks that follow.

To see a video clip of earthworm observation go to the companion website for the book at **https://study.sagepub.com/chambersandsouter**

Responding to the environment

Animals will respond to the environment in a variety of ways (e.g. by moving towards or away from a stimulus). They can avoid predators or conflict with competitors by using their senses to detect their presence and responding appropriately. Finding a mate frequently involves responding to visual, auditory, and chemical attractants. Different animals are active during the day and at night time, and they must have means of detecting, coordinating and responding to environmental cues. Similarly seasonal responses, including hibernation and migration, are triggered by environmental cues (e.g. day length and temperature). However they must respond in a coordinated manner in order to prevent random responses and ensure survival. The sensory system detects physiological and environmental changes and the whole organism responds to that information.

ACTIVITY

Responses

1   Investigate food preferences by observing a birdtable (if direct observations are made the children are to be aware of the need to remain still and quiet). Food trays can be examined and recorded at regular intervals; indirect records via video can provide useful opportunities for observing specific behaviours (e.g. predation and competition) however secondhand experiences are less engaging.

2   Nesting boxes also give opportunities for direct observations or video records and counting the numbers of visits by parent birds.

3   Think–pair–share why birds build nests.

4   Predation can be modelled in a variety of ways. Grazing effectiveness can be replicated with the children, as 'predators' and tricolour spirelli as 'prey' within the simulation. Scatter 2kg of the pasta in a marked area of the school grounds measuring 15x15m. Provide containers (e.g. yoghurt cartons, margarine tubs) for the 'predators' to collect their 'prey'. Run a series of tests to find out how effectively they 'graze' during one minute; what are the proportions by colour; how does this change during each subsequent period of grazing? The records can be presented in a variety of ways (e.g. single-coloured pasta piles or tally charts illustrating the collections). Reflect on the capture rates of the prey, the numbers

retrieved on the 2nd and 3rd grazing, and the increasing competition pressure on the predators as the trials progress.

5 Observe and record animal movement (e.g. a snail observed through a Perspex sheet, the location and distribution of brine shrimp swimming in a tank, stick insects walking across a sheet of paper etc.). Video recordings can support these experiences by allowing learners to access play/replay/pause and slow motion.

Citizen science

The public, as non-scientists, have the skills, abilities and willingness to contribute to the collection of data that informs scientific research in a wide range of projects.

Nature's Calendar (www.naturescalendar.org.uk/survey/)

This is an excellent example of 'citizen science' in the UK. It provides everyone, including children, with an opportunity to contribute to the establishment of meaningful scientific data by observing seasonal events in their neighbourhood. Observation and recording each event is well supported, and the data are shedding light on climate change:

'Nature's Calendar is the home for thousands of volunteers who record the signs of the seasons where they live.

It could mean noting the first ladybird or swallow seen in your garden in spring, or the first blackberry in your local wood in autumn.

You don't have to be an expert to take part.

Lots of help is given, including a free downloadable nature identification booklet.

This kind of recording has moved from being a leisure hobby to a crucial source of evidence as to how our wildlife is responding to climate change.

Phenology is the study of the times of recurring natural phenomena, especially in relation to climate.

It is recording when you heard the first cuckoo or saw the blackthorn blossom. This can then be compared with other records.'

Woodland Trust (2017)

The Big Garden Birdwatch

The RSPB's citizen science project has been carried out in the UK each January since 1979. It sheds light on the various population changes, increases and decreases in the UK's bird populations.

> ## ACTIVITY
>
> **Citizen science**
>
> Enrol in a citizen science project and contribute recordings to the dataset.

Detecting the environment

All living things are capable of responding to stimuli from their environment. Even the simplest ones are capable of detecting what is going on around them – gravity, light and chemicals provide abundant insights into their surroundings and information that will help their survival by responding to the availability of food and the presence of harmful chemicals in particular. Multicellular organisms frequently possess specialised cells, tissues and organs that collect information from the environment. This information is processed in a coordinated way to the entire organism. Plant responses in relation to photosynthesis are driven by light. They are also able to show a variety of growth and physiological responses to daily and seasonal variation, temperature changes, a host of internal and environmental chemical stimuli, touch, and of course the capacity to regulate growth in relation to gravity. Animals, in turn, are capable of responding in a variety of ways to the same range of variables. Movement is the most obvious of the responses which allows them to exploit their environment and avoid harm. It is important to note that human senses have limitations to their accuracy and will vary from person to person.

> ## ACTIVITY
>
> **Senses**
>
> Encourage learners to think about what they detect from their surroundings. Think–pair–share the senses they use during different scenarios (e.g. walking to school, playing a ball game, shopping for sweets, getting ready for bed, eating dinner).

Human sense organs

Specialist organs support tissues that have evolved to detect the environment by converting various forms of energy into electrochemical signals that are then transferred via the nervous system. The brain receives information from external and internal sources and acts to control and coordinate an animal's survival. These organs include:

- the eyes which detect shape, movement and colour
- the ears which detect sound, movement and gravity
- the tongue which detects dissolved chemicals
- the nose which detects gaseous chemicals
- the skin which detects hot, cold, touch and pain

ACTIVITY

Human sense organs

1   Devise an outdoor or indoor trail that triggers a variety of senses.
2   Play and learn the '5 senses song' from YouTube.

The eyes and vision

The external structure of the eye is designed to protect it. The eye's internal organisation ensures that light is focused on the retina – the light-sensitive tissue that passes optical information on to the brain.

The eyes are well protected from impact by being located inside large bony sockets in the skull. The blinking reflex prevents objects entering the front of the eye as do lashes and brows. Tears are a miraculous fluid that lubricate the eyes and the lids; moreover they prevent infection with salt and antibacterial enzymes that are concentrated within each tear.

Three sets of muscles are significant; firstly the muscles that lie outside the eye control its movement; secondly the iris is a muscular structure that controls the entry of light through the pupil and into the eye; and thirdly the final group of muscles contract to change the shape of the lens and consequently focusing. The bulk of the interior of the eye is made up of a gel that maintains its shape in conjunction with liquid at pressure. The retina is the tissue that is responsible for trapping light, and then converting this into nervous messages that travel along the optic nerve to the brain.

ACTIVITY

The eye

1   It is possible to observe anatomical features and infer the ways in which the eye works by carrying out some simple tasks. Ask learners to sit straight up, face the same way, and without moving their heads to:

(Continued)

(Continued)

look at the ceiling,

look at the floor,

look out the window,

look at the door.

They might repeat this activity in pairs, acting as observer and recorder in turns. Can the children infer the presence of muscles around the eye? Prompts related to movement and the skeleton might help here.

2 For this activity reduced ambient lighting is required, but a blackout and curtains are not normally necessary since this will work in a normal classroom on a dull day. Ask the children to sit in pairs to face each other and make observations about their eyes in order to describe various features, including the lids, lashes, brows, whites, irises (coloured part), and pupils (black part). Ask them to watch the pupil carefully when you flip the classroom lights on and off quickly. The activity is designed to show the pupillary reflex that protects the eye from receiving too much light, and to seek inferences that since the iris is moving this must involve muscles.

3 Invite the children to close their eyes and *very* gently place their index fingers on the upper lids and describe what they feel as they rotate their fingers gently over the lid against the eyeball. Can they describe it as being slippery? Are they able to infer that the function of tears is lubrication?

4 Research the advantages of tears tasting salty.

5 Ask the children to hold an index finger, pointing upwards, about 30cm in front of their nose. Invite them alternately to look at their finger (close vision) and without moving the finger to look at an object on the classroom wall or outside (distant vision). Do they notice the changes that take place in focusing? When they are looking at their finger it is in focus and the distant objects are out of focus and vice versa. Are they able to infer these changes are brought about by internal eye muscles?

6 Use different mirrors, spoons etc. to show the optical effects of reflective surfaces.

7 Prepare four or five short paragraphs of text, print these in different colours, and then use coloured acetate sheets, filters, spectacles and torches. Can they read all the paragraphs?

8 Experiment with darkness by building tunnels under tables or turning lights off in a darkened room.

9 Children may be willing to share their experiences of eye testing at the optician's. Simple visual acuity tests may be carried out with a standard Snellen's chart or

those modified for pre-reading stages. Alternatively they may design their own visual-testing method; groups could look at different properties – do they see equally well with each eye? (This tests binocular vision and colour vision.)

10 Highlight their dependency on sight. When they are blindfolded can they recognise a friend by touching their face or clothes? Can they walk in a straight line while blindfolded? Discuss their feelings following this simulated blindness.

The ears, hearing and balance

Sounds are produced by vibrating objects. They travel from their source by causing the air to vibrate and are directed towards sensory tissues by the external ear. Vibrating air travels down the ear canal, which is protected by outward pointing hairs and wax, towards the eardrum, which in turn vibrates, passing the sound energy on to the middle and finally inner ear, where it is converted into nervous impulses that are carried to the brain. The inner ear also plays a very important role in maintaining our balance. The movement of fluid inside the semi-circular canals which lie at right angles to each other informs the brain about our three-dimensional position – up/down, back/forward, left/right.

ACTIVITY

Sounds

1 Can children feel the vibrations while they talk with their lips touching a balloon?

2 Make a sound by 'twanging' a ruler against the tabletop. Make the note change (can you make a higher or lower pitch and volume?).

3 Give the children a balloon and set the challenge, 'How many different sounds can you make without destroying this balloon?' There are of course lots of ways to do this and this can be quite an inventive activity. (You might place rice or peas inside a balloon, inflate it, and then make some interesting sounds.) Sound is made in musical instruments by 'Bang, Twang, Blow, or Bow' – a useful classification.

4 Can learners make 'music' with objects in the classroom?

5 'Foley' artists recreate sounds in the movie industry. Invite groups to make a story, support it with sound effects that match the scenes in the story, and record it as a podcast. Several audio recording apps are available including Audacity and Garage Band.

(Continued)

(Continued)

6   'Animal Sounds', a song by A.J. Jenkins, describes the sounds that animals make. Lots of sound files are available on the Internet and can be used to celebrate the extraordinary variety of sound communication that is produced within the animal kingdom.

7   How is their hearing influenced when learners cup their hands in front of their ears or behind them? Does it help their hearing to put a cone of photocopy paper around the ear? (Historically ear trumpets helped those with impaired hearing; here you can refer to the mobility of dogs' and cats' ears when they are exploring their environment.)

8   Provide opportunities for children to experience silence, or near silence, and recognise the contrast between silence and noise. Step outside the classroom and carry out a sound survey. It helps if the children close their eyes and concentrate on everything that they can hear during the short time span (e.g. 20–30 seconds). Animal sounds and man-made ones will dominate, although it is likely that the children might also hear elves, bears or monsters in the undergrowth!

9   Are the children able to identify the direction of sounds when they are blindfolded? Test this in the classroom by opening the door or a desk drawer for example, or turning on a light or a tap, scratching on a desktop etc.

10 Investigate sound proofing with a range of different materials.

11 'It was so quiet you could hear a pin drop'. Investigate how far away they can hear a pin drop. Ensure fair testing and that a standard 'pin-dropping' protocol is established. We suggest that coloured paper clips are used for this.

12 Encourage the children to classify different sounds.

13 'Don't put anything smaller than your elbow in your ear'. Discuss this adage and its importance with the children.

14 Can they describe what it feels like to be dizzy? What made them feel dizzy? How long did it last?

The skin and sensitivity

Human skin is a large organ and is much more than a body covering to prevent the entry of pathogens. The skin carries out a diverse range of functions:

- Our skin prevents water loss; dehydration is the major risk for burns victims.
- Skin manufactures vitamin D in sunlight.
- It plays a significant role in temperature control so that when we are cold we shiver, blood moves away from the skin and our body hair stands up, leading to

goose pimples, and when we are hot the skin releases sweat and blood moves to the surface to radiate the surplus heat.

- Many microorganisms live on the surface of the skin. These prevent the establishment of pathogens.

Sensory nerve endings extend into the skin. These react to a variety of stimuli, including heat, cold, touch, pressure, vibration and injury. These nerve endings are unevenly distributed around the skin, with the highest concentration on the fingertips, lips and genitalia.

ACTIVITY

Skin

- Select a range of resources which offer a variety of textures that the pupils may explore to make observations and descriptions. These can be presented in 'feely bags'. The collection should include the widest range of materials (e.g. feathers, nails, sandpaper, wood, plastic, fabric, dried pasta, jelly, yoghurt, dough etc.). Learners can extend this activity by describing their feelings when they walk over different surfaces in their bare feet.
- Can they identify familiar objects that are wrapped inside containers?
- Prepare 3cm squares of different grades of sandpaper. Can blindfolded children place these in their correct order by feeling the sandpaper? How successful are they when they touch the paper with different parts of their body (e.g. their elbow or chin)?
- Discuss today's weather. Encourage a full discussion and invite learners' speculation about which part of the body they are using to collect evidence and make their comments. Many of the properties they describe, and the weather they feel, will have been detected by the skin.

Taste and smell – the related senses

Gustation and olfaction (i.e. tasting and smelling) are interrelated senses that respond to chemicals but are combined with strong visual signals to identify flavours and warning odours. Taste is detected when dissolved chemicals react with receptor cells in the mouth. Visible lumps in the tongue (papillae) include hundreds of taste buds and these are also found all around the mouth. Taste buds respond differentially to groups of chemicals which are described as being sweet, sour, salty, bitter or *umami* – a Japanese word that means 'pleasant savoury taste'. Olfaction (i.e. smell) is detected by nerve endings in the roof of the nose which combine with food molecules to send messages to the brain. The chemicals that are detected by the olfactory system are not readily classified.

ACTIVITY

Taste and smell

There are significant safety issues associated with these senses so ensure that none of the materials will promote food allergies and that large pieces of food are avoided. Powder should not be sniffed directly, and food containers with holes punched in the lid will prevent inhalation and possible irritation. Hygienic practices should be followed during all the activities.

1 Identify foods through blind tasting (e.g. fruit, potato crisp flavours, confectionery and vegetables).
2 Make labels with words that describe the flavours of a range of foods.
3 Describe odours from a variety of distinctive locations (e.g. deciduous or coniferous woodland, the swimming pool, a carpenter's workshop, a cattle shed etc.).
4 Describe odours from a variety of products (e.g. vinegar, toothpaste, perfume, orange peel, essential oils, honey, curry powder, coffee powder etc.).
5 Reflect on the smells that are released by good and bad food/well cooked and burnt food.

Summary

Animal behaviour depends on the capacity of an organism to be able to detect, organise and respond to environmental stimuli. One vital characteristic of all living things is that they can achieve this. While responses in animals might be more obvious, those in plants are significant, and even those that take place in the most primitive microorganisms assist their survival and are indicative of their sensitivity and responsiveness. Exploring their own senses provides children with an initial insight into their extent as well as their limitations. These observations can be extended through consideration of the range of sensory input. Insights into the ways in which their sensory organs work can similarly be extended while incorporating related targets in health and well-being. Children's and animals' behaviour and learning can be introduced and supported with a range of purposeful, first-hand and direct experiences of living things within a classroom setting. Such experiences, while extending their sense of wonder of living things, can also contribute to promoting responsible practice in taking care of living things.

REFLECTION POINTS

- Consider intelligence, multiple intelligence, and animal intelligence. Think also about the context of learning in both humans and animals.
- Make a list of the different forms of communication that animals are able to undertake. Research further examples of each of these.
- Why are 'citizen science' projects important? (What's in it for science? What's in it for the individual? What learning might take place?)
- What opportunities are available for you and your class to take part in local/national projects?
- Compare and contrast the ways in which plants and animals respond to the environment.

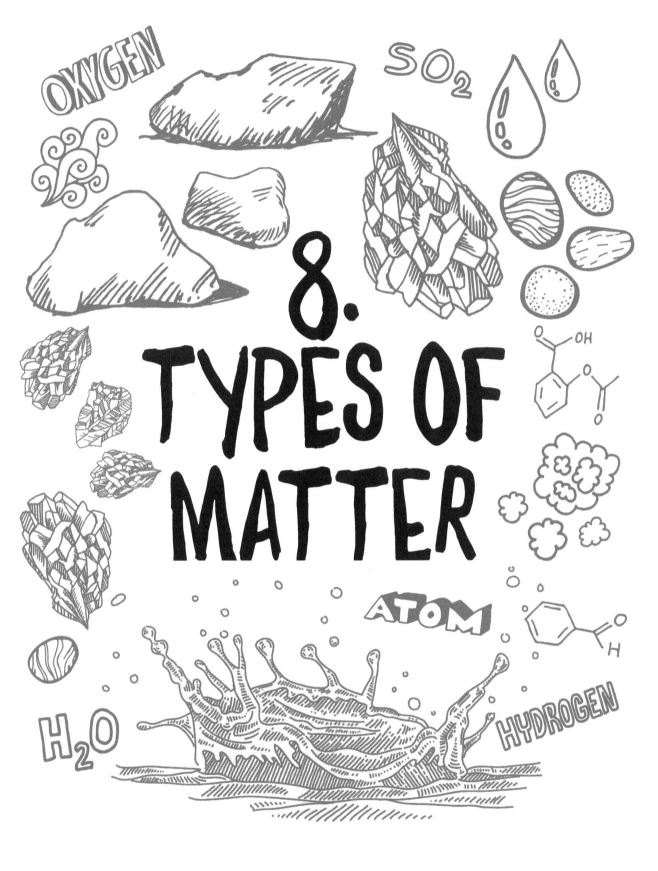

8. TYPES OF MATTER

Learning Objectives

Having read this chapter you should:

- have a good understanding of what matter is
- have a secure knowledge of what is meant by a solid, a liquid and a gas
- understand the processes by which matter can change state
- identify situations where matter can expand and contract due to temperature change

Overview

In children's terms, matter is the coverall or general term for everything that is around them. In a classroom the chairs, tables, pens, paper, paint etc. are all composed of matter. Matter can be described as anything that has mass and occupies space. A tree for example would be said to be composed of matter as it both occupies space and has mass. A precise definition of what space is and what mass is not easy for young children, and it is suggested that initial teaching in this area focuses around discussion of pupils identifying and describing objects in terms of mass and space. Balloons for example can occupy a large space but have a small mass whereas a piece of lead will occupy a small space and have a large mass. In introducing children to the types and states of matter, presenting them with a range of simple everyday materials and then allowing them to describe or write about what they see, feel or even think about materials is a very good starting point, and will allow you to determine a baseline for children's knowledge in this area.

States of matter

For primary school children it is sufficient for them to appreciate that there are three states of matter: *solid, liquid* and *gas*. All materials in their world can be categorised in terms of these three states. Stars, however, are a state of matter known as plasma. This is a state of matter that only exists at very high temperatures or energies, and is not really appropriate for discussion with young children. Interestingly, as most of the matter in the universe is associated with stars, it is the most common state of matter in the universe. Our normal solids, liquids and gases etc. account for only a small percentage of all the matter in the universe. Plasma screens excite small amounts of glass to high energies and in effect become 'plasma', but the explanation of this is complex.

Solids

Solids are materials that have a definite shape, size and volume. That shape will not change unless some external action occurs (i.e. you press or squeeze it). In general, if you place a solid down on a surface it will not flow or deform.

Some solids don't easily fit into this definition however. Soft solids may appear to flow (e.g. hair is solid yet people can run their fingers through it). Children can be excused for thinking it is not solid because it is soft, can fold over, and even hang over the side of a table for example. It is a solid because its volume has not changed and it has not separated and then somehow rejoined.

The same is true of fine, particulate solids like sand. When there is a lot of sand it will be able to be poured and take up the shape of a container, but each individual grain will remain in the same shape with the same size.

In general, solids will stay in one place and can be held. They do not flow or spread out in individual terms. We can even slice and cut them.

Descriptions similar to the ones above are perfectly acceptable from young people as examples of what makes a solid a solid.

To be more technical, solids have a regular structure (i.e. the particles they are composed of are arranged in such a way that they are close together and have a regular arrangement, which repeats itself for a very large number of particles).

The particles are also closely packed together and this makes it very difficult to squeeze or compress a solid. The particles (atoms or molecules) are held close together by chemical bonds and will remain this way unless acted upon by some

Solid

Figure 8.1

Designua © 123RF.com

external force or energy. A coin for example cannot be compressed. We can deform it but not 'squeeze' it into a smaller volume. The same is true for glass or plastic or a pebble, say.

A sponge for example *can* be compressed, but it is the air within the sponge that is being expelled or compressed. The particles of the sponge are not being compressed.

In solids, the particles are not fixed in that they can vibrate in their position. They cannot move about freely but they can vibrate. If we heat a solid they can vibrate more, and this can lead to an increase in temperature and also volume. The added energy increases the volume of the solid due to the particles vibrating more and pushing apart to some extent. This is the cause of materials expanding. This is noticeable in buildings for example. When the sun shines strongly on a window for example, the glass and frame heat up and expand slightly. This expansion means we can hear small clicks and noises as the materials gradually push apart and change position slightly. It is also noticeable in roads, railways and bridges. When heated, concrete and steel will expand and can cause roads and bridges to buckle or deform.

Photo 8.1

By Lankyrider (own work) [CC BY-SA 4.0 (http://creativecommons.org/licenses/by-sa/4.0)], via Wikimedia Commons

The photo shows a particularly extreme example of a long rail which was laid in the hot sun without allowance being made for its expansion.

We can counter this by leaving joints in the road and rails allowing them to expand when heated. In the next photo the road expands in hot weather and the teeth move closer together. The converse happens in cold weather where the sections of the road contract and the teeth separate more.

Interestingly, the supersonic airliner Concorde used to expand by almost 30cm when flying at twice the speed of sound. The fuselage and paint had to be specially designed to allow for this expansion and the later contraction during landing.

Photo 8.2

The opposite also happens when solids are cooled. When cooled, the molecules vibrate less and they do the opposite of expand and contract. These effects are not easily noticeable with small amounts or lengths of solids, but with larger sizes the effects can be easily seen.

Practical expansion

The effects of expansion and contraction are also easily observed at home. Doors and windows can become loose or sticky when opening and closing in different seasons. Metal gates in summertime may be more difficult to close as they will expand and not meet exactly at the centre. Many bridges will have one end of the bridge fixed to a concrete base for example, but the other end will be on a roller mechanism which allows the bridge to expand and contract as the temperature varies. You can see these if you look closely at the supports.

Liquids

Liquids are commonplace and children will all be familiar with what a liquid is. They bathe in, swim in, and drink water. They will also have experience of milk, coloured juices, oils and such as they will have been seen in containers around school and the home. Liquids have a fixed volume but not a fixed shape. They will fill or take up the shape of any container they are poured into. Their chemical composition means that when poured into a container they will fill up to a certain level. The liquid can move freely around the container and the particles of the liquid are free to flow over and around each other. The bonds holding the particles together are weak but still act on each other. For example, if you spill some water onto a table the water will spread out in little drops. These drops are held together by the attraction of the particles but they are weak, so the drops do not form a large body of water and instead will gradually disperse. When you pour water it also forms a little stream or group of droplets that are weakly held together. The forces holding the particles together are weak but their effects can be seen. As mentioned earlier the volume of the liquid will not change unless you add or remove some of it, but this may not be obvious to learners. Pouring a liquid from one container to another can cause some difficulties for children in accepting that the same volume is in each. Some will believe a tall thin container holds more of a liquid than a small flattish one.

Liquid

Figure 8.2

Designua © 123RF.com

As the bonds are weaker than in solids the particles can flow over and around each other. If we place our finger in a cup of water for example, the particles are pushed to either side and our finger feels little or no resistance when doing this.

A good activity regarding the volume of a liquid is to pour 250ml of water into a cup, then to pour it into a bottle, then into a bowl, and then back into its original container. It will still have a volume of 250ml. As said earlier, a tall thin beaker may appear to contain more liquid than a wide flat bowl. Learners can determine for themselves that the volume of the liquid is not solely dependent upon the height it reaches in the container.

In a container the liquid's surface will be flat and level. The nature of liquids is that we cannot have a point on the surface of a liquid higher than any other point on the surface. As the particles are free to flow, a liquid particle at a high position can 'fall' or flow to a lower position. The particle falls due to gravity acting upon it.

If we place an object into a container of water, the water level will rise due to the object displacing the liquid so that it has to go elsewhere. The liquid moves to the side and displaces other particles. At the surface the water will be level.

Even although the particles are relatively free to move in a liquid they are still in close proximity to other particles. This makes it very difficult to 'squeeze' or compress water. A filled syringe of water cannot be compressed easily.

This incompressibility of liquids is useful and is the basis of what we refer to as hydraulics. Hydraulic devices work on the principal that if you squeeze a cylinder with fluid in it, that fluid will be pushed out and along a tube to the far end of the device. The far end could be another cylinder, and by squeezing the first cylinder this ultimately moves the second cylinder. If we arrange the diameters of the cylinders in the correct way, we can apply a stronger force to the second cylinder. This is how hydraulic car jacks, car braking systems and earth-moving machinery are able to move large loads or apply great forces. For example, when you press your foot down on the brake pedal in a car force is applied via a hydraulic pipe to the brake pads, and these are then pushed against the discs which 'grip' the wheels, thus slowing the car.

Liquids and temperature

Liquids are affected by changes in temperature in much the same way as solids. They expand when heated and contract when cooled. This seems fairly obvious, however it has some important applications.

Thermometers

Thermometers are used to measure the temperature of objects. The simplest form of thermometer is possibly the liquid in glass thermometer. A small amount of liquid is placed in a reservoir and then connected to a thin tube.

If we heat the liquid in the reservoir, the liquid expands and 'flows' up the thin tube. As the temperature increases, the expansion of the liquid increases, and it flows further up the tube. If we mark a scale on the side of the tube we have a working thermometer.

Mercury thermometers were once very commonplace. Mercury's expansion is constant and its silver colour enables us to see it clearly as it moves up the scale. Health concerns regarding spilt mercury or broken thermometers have now resulted in its removal, and glass thermometers may have alcohol or some other liquid as the expansion material. Modern thermometers are now often electronic devices. Liquid expansion thermometers however are relatively simple to make.

Photo 8.3

iStock.com/ru3apr

To see video clips of expanding bars and liquid/gas thermometers go to the companion website for this book at **https://study.sagepub.com/chambersandsouter**

This exemplifies the behaviour of liquids in varying temperatures. Occasionally the expansion of liquids has caused a liquid to overflow its container. If a motorist fills their car with petrol (which is kept at a constant, cool temperature) and then parks it in direct sunlight the petrol can heat up and expand. If the petrol tank is full, the petrol will expand and pour out of the tank via the filler opening. This leads to a spillage (and wastage) of fuel for the driver.

Gases

Gases are the third state of matter, and as mentioned before children will be very comfortable discussing and describing their thoughts and ideas with regard to gases. Air is the most common gas they will be aware of and they will have direct experience of winds and papers being blown away etc. Air is a mixture of gases, mainly nitrogen (78%), oxygen (21%), argon (1%), water vapour (< 1%) and carbon dioxide (0.03%). There are very small amounts of other gases but these are not relevant in this context. Carbon dioxide actually makes up a very small portion of air, but as we exhale carbon dioxide it helps children appreciate this is in the atmosphere.

Particles in a gas are widely spread with very little attraction between them. They are in a fairly energetic state and move about at high speeds. They have no fixed shape or volume.

Gas

Figure 8.3

Designua © 123RF.com

If we have a container of gas and remove the lid, the gas will move and spread around the room or area in which it has been released. Due to the particles' motion and little or no attraction, the gas will diffuse and continue to spread. Ultimately it will spread so widely that it will appear to have gone. If we spray perfume on an area other people in the room will smell it as the perfume gas spreads and gradually occupies the space it is in. We can smell the gases being given off when someone is cooking or the smell from a food outlet some distance away.

As the particles in a gas are widely spread, unlike solids and liquids, gases can be compressed. We can insert a large mass of gas into a fairly small container, but as we continually add more gas the pressure inside the container increases. Most pressurised containers are made of some form of steel. SCUBA (Self Contained Underwater Breathing Apparatus) divers strap large canisters of compressed air to their backs, and this is then gradually released to allow them to breathe underwater.

The air at high pressure has to be released at a lower pressure to allow a diver to breathe easily. The regulator connected to the mouthpiece matches the pressure that the diver experiences due to being underwater. It then allows air from the cylinder (at very high pressure) to be at a suitable pressure to allow the diver to breathe. This is very important. The pressure from the water surrounding the diver makes it difficult for them to breathe in, and the mouthpiece helps by releasing gas at a high enough pressure to help the lungs expand and fill with air. If we took a long tube underwater and attempted to use that tube as an air supply it would be very difficult, if not impossible, to do so at very small depths. The pressure of the water surrounding the diver makes it almost impossible for them to breathe in correctly at depths equivalent to the shallow end of a swimming pool.

It is relatively simple to show that gases are compressible. Use a syringe or bicycle pump. Close one end with your finger and press the plunger or pump. This will resist as it gets towards the end, but when you release it it will gradually move back towards its original position.

The explanation for this is that originally the air within the tube was at a normal air pressure. As you compressed the tube the air was forced together and as a result the air pressure was increasing. This was the resistance you may have felt as the plunger was depressed.

When you released the plunger the gas in the tube was at a greater pressure than the air outside, so it gradually forced the piston back out.

Expansion and contraction

Gases are also affected by changes in temperature. When a gas is heated the particles become more energetic and move more quickly, causing expansion. A simple experiment to show this is to inflate a balloon and tie it to a stand. Using a cloth tape, measure its circumference. Gradually heat the balloon with a hair dryer for a couple of minutes and measure its circumference again. You will find that the balloon has expanded because the air inside it has expanded. If you place the balloon outside for a couple of minutes and measure again you will note it has reduced in size.

An interesting example of this was observed by the author during an in-service event. A helium balloon was being used for an experiment and it was left outside, tied to a weight. (It was a cold January morning.) When the author came back out it was wrinkled and deflated, as if it had burst. The author took it indoors and within minutes it was back to its fully inflated shape. The difference in temperature going from inside to outside was enough to cause a large decrease in volume.

Most gases are transparent (i.e. we cannot see them) but some gases have a colour to them. Many gases have an odour however, and this can be used to describe the behaviour of gases to learners.

Changing state – melting

It is relatively easy to change substances from one state to another: solid to liquid (melting ice) or liquid to gas (water vapour from a kettle) are quite commonly observed in everyday life.

When changing state, the particles themselves do not change or combine with anything. It is their structure that changes. Their interaction and attraction to other particles change, generally due to the addition or removal of energy from the material.

In a solid, the particles have strong attraction to the surrounding particles. This causes them to be held together in some form of row or column. When heat energy is applied to the particles they vibrate more and more. Initially this vibration leads to expansion of the material, but further heating gives rise to a change in the overall structure of the particles. With enough energy the particles will break the tight structure and the material becomes a liquid. The particles are still loosely attracted in that they can move around, but they are not held in a regular structure.

Interestingly the vibration of the particles is an indication of the temperature of the material. The greater the vibrations, the higher the temperature. In scientific terms the average kinetic energy is often referred to as a measure of the temperature. This is beyond the scope of young children, but this introduction to moving particles, even in a solid, and this movement being used to explain physical phenomena, are a good grounding for further work.

Most materials will melt and change state when heated but not all. A common question from inquisitive children is

'Can you melt wood?'

The answer is no. This is due to the composition of wood itself. It is a combination of cellulose, water, and other organic materials. When heated these chemicals decompose and oxidise, leading to combustion, and this destroys the structure of the wood itself. This oxidation changes the wood into charcoal (carbon), methanol, and other products.

The temperature at which a solid changes to a liquid is referred to as its melting point, and this can be used to identify certain unknown materials. We can measure the temperature at which it melts to give us an indication of what the material is.

Table 8.1

Chemical	Melting point in °C
Hydrogen	−259
Mercury	−39
Potassium	64
Tin	232
Aluminium	660
Copper	1083
Nickel	1452
Tungsten	3399

As you can see in Table 8.1, the melting points of materials vary widely. We tend to identify substances as solid, liquid and gas, but only at room temperature (generally around 20°C). Metallurgists are people who specialise in designing metals so they are suitable for our purposes (e.g. structural steel for buildings is designed to be relatively strong but does not need to withstand high temperature, whereas metals used in aeroplane engines need to be strong, lightweight, and able to withstand high pressures and temperatures).

Children's experiences of melting will probably be based around ice melting to become water. Ice melts at 0°C. It is not a coincidence that this is the value. It was *chosen* as the basis for the temperature scale. The scale relies on two fairly constant temperatures:

• The temperature when ice melts (define this as 0°C).
• The temperature when water boils (define this as 100°C).

Thermometers use these two points as the basis for their scale, and divide the distance on the glass into 100 sections, which then gives us a temperature reading.

Changing state – freezing

When a liquid freezes it undergoes the opposite process to melting. The temperature around the liquid is reduced and heat energy is transferred from the material to its surroundings. The average energy of the particles of the material falls, as does its vibrations, and it reverts to its more ordered structure (i.e. a solid). This temperature is the same as that when the object melts. It is a reversible process and can repeat itself again and again.

When materials change from solid to liquid, or liquid to solid, they don't really change their volume. There are occasionally slight differences, in water and ice for example (ice expands slightly), but in the main the volume remains pretty much constant.

Changing state – boiling

Vapourisation is the name given to the process where a liquid changes state to a gas. In a similar way to melting, the particles need to gain more energy to overcome the attraction of the other particles, and in the case of liquid to gas, to be able to 'break free' from the other particles. To break free the particles generally need to gain a great deal of energy, and in order to do this will need to reach a high temperature. In water, this is referred to as the boiling point which is generally 100°C. At this temperature the water molecules can break away from the other molecules and become gaseous.

When boiling water, the water at the top of a pot or kettle for example can turn into steam and then spread through the air. The water at the bottom turns into a gas (i.e. a bubble) and this then rises to the top. This is why boiling water bubbles.

When material changes state from liquid to gas there is a great increase in volume (by about a thousand times). One cubic centimetre of liquid will become approximately 1000 cubic centimetres of gas.

When boiling liquids great care must be taken to ensure the increase in the volume of the gas does not increase the pressure in the container too much. If the pressure inside the container becomes too high, the gas inside can literally burst the container due to this increase in pressure.

The opposite also can occur. Fill a container with water vapour for example, then close the lid and allow it to cool. When the water vapour reverts to liquid water, the pressure in the can will decrease and the air pressure outside will crush the can.

Changing state – condensing

Condensation is the opposite process to boiling. It is the process of a gas changing state to a liquid. On a cold day the windows in a house will be colder than the air in the house. Any water vapour in the house may condense on the cold window to form what is commonly called condensation. It can also happen on walls that may not have been constructed properly, and as a result can become cold on the inside of the house. This is very unhealthy and mould can form on these damp patches of 'condensation'.

This is easily observed by leaving a cup or beaker of hot water by the window. The water vapour from the beaker spreads around, and when it encounters the cold surface reverts to a liquid and forms droplets on the window.

On cold days, the water vapour from our breath condenses when it makes contact with the outside air and we can see the vapour when we exhale.

Clouds are examples of water vapour. Very tiny droplets of water condense and are just large enough be seen. These droplets are so small that they fall at an extremely slow speed. They generally form in air that is rising, and the combination of the air rising and the particles having a very small 'dropping' speed means the clouds appear to remain in the air or 'float'. When a plane flies through clouds the water particles come into contact with the plane and it wobbles a little.

We can do another experiment which shows condensation by filling a beaker with ice. Water vapour in the room condenses on the beaker and the outside of it becomes wet. This shows condensation occurring, but remember also that air contains water vapour as one of its small but important constituents.

Steam

When water is in the form of steam, it is at a temperature greater than 100°C. If the steam comes into contact with a surface at room temperature, it will condense to water and then cool to the temperature of the room. The steam releases a great deal of energy when it condenses and also as it cools down. This can be used to dislodge heavy stains. The undersides of cars are steam cleaned to help dislodge oil, tar, hardened grease etc. Steam can be also used in homes to clean surfaces, but it is very dangerous if it comes into contact with skin. A lot of energy is released when it condenses and cools and this can cause serious burns.

The processes involved in changing states is summarised in Figure 8.4.

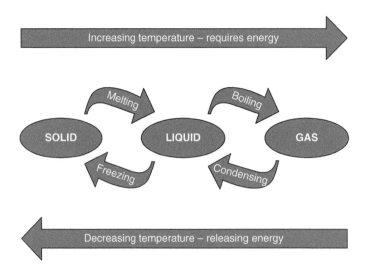

Figure 8.4

Evaporation

Evaporation also occurs when a liquid changes state to a gas. It is *not* boiling however, as it occurs at temperatures below the boiling point of the liquid. Evaporation is common in everyday life (e.g. we hang out our washing to dry, puddles dry up, our hair dries naturally etc.). The process is similar to boiling in that the particles need enough energy in order to turn into a gas. At low temperatures they would appear not to have enough energy. Consider a layer of water as shown in Figure 8.5.

Figure 8.5

The particles at the top of the drop can receive more energy than those at the bottom from the surroundings. Ultimately they can obtain enough energy to turn into a gas and leave the surface. If there are additional sources of energy such as sunlight and wind say, more particles will eventually obtain enough energy and gradually the whole drop will evaporate. When we hang out washing, it will dry more quickly if there is sunlight and a reasonable breeze. Both the wind and the sunlight supply energy to the particles to allow them to evaporate.

Additionally, the particles that do evaporate also extract or take heat from neighbouring surfaces, leaving them slightly colder. This is why we sweat when we become warm.

Our body regulates our temperature naturally, so if we are in a warm environment or are exercising, which increases our temperature, our body produces beads of sweat on our skin. These beads will evaporate, taking heat from their surroundings, including our skin, and therefore cooling us down.

This is also why we feel colder when it is wet and windy. Rain on our arms for example is colder than us, and also will evaporate quickly in a strong wind. This takes more heat from our bodies and we become very cold. A simple experiment that shows this is to spray some water on the back of your hand and then blow or fan some air over it. You will feel your hand becoming colder.

Perfumes rely on evaporation. Liquid perfume is placed on your arm and it evaporates, turning into a gas. That gas then spreads into the air and people around you can smell it. A problem with perfume is that it can evaporate too quickly and the fragrance will thus not last for any length of time. Perfume companies are aware of this, and along with the chemicals that cause the desired 'scent', there are other chemicals such as oils which delay its evaporation.

Summary

This chapter has given an overview of materials and their properties. It has gone from a relatively straightforward description of how we may classify materials to a fairly detailed description of the molecular processes occurring when materials change state. It was written with the curriculum for young children in mind, but also sought to increase teacher confidence. As such it will have gone beyond what you are teaching but will allow you to answer questions related to topics beyond the scope of the curriculum. Some of the content overlaps with other chapters, but this is a common occurrence within the text, as many topics are overarching and do not fit into one particular 'box'.

REFLECTION POINTS

The following activities are designed to allow you to use your knowledge of matter and its properties to inform your teaching.

- How would you explain to learners that a liquid can change to a gas even when it is cold?

  - Use examples relating to their own experiences (e.g. their hair drying 'naturally' even when they are in a room which is not warm or if they go out when it is still wet).
  - Ask how a pavement dries after rain, even in winter. Explain this using information like the cold wind giving energy to the water at the top of the puddle and this small section of water being able to evaporate.

- What material should a coin be made of? What are the properties that we would want if for example we were to make a new £5 coin?

  - The answer would involve establishing the properties we want in a coin. It must feel heavy (perhaps) so we know it is in our pockets. It should be hard-wearing as it is jangled together with lots of other coins in tills and in pockets. It should be shiny to give the appearance of value. It must not rust or tarnish so that it keeps its look.

- How would you explain why wood or timber is so widely used for building homes and furniture?

  - Your answer would involve examining wood's properties. It is freely available as trees are found throughout the country. It is relatively lightweight and can be transported from a forest to where it is required fairly easily. It is strong for its size and people's weight can be supported on relatively small sections. It can be sawn, planed and sanded to make it into the shape we want it to be. It looks good.

- Describe a solid, a liquid and a gas in simple terms.

  - A solid is a material that does not change its shape or size. If you move a stone say from the ground to a bench, it remains in the same shape as it was when you picked it up.

○ A liquid is a material that does not change its size but does change shape. Pour 100ml of water from a bottle into a cup and it is still 100ml of water. Its shape is now that of the cup.

○ A gas has no fixed size or shape. If we had a balloon filled with helium and opened it, the helium would disperse and spread around the room. The gas released when paint dries makes its smell noticeable initially, but after a while it gradually spreads until it is no longer apparent.

The answers to these reflective questions rely not only on your own knowledge of the properties of matter for example, but also on being able to suggest a range of answers you could give to learners and allow them to explore or investigate further. Furthermore, terms may have a 'normal' meaning but also a slightly more specific scientific meaning which for young people may have to be gradually worked on. Young children may think a solid is hard and strong, but cotton wool is a solid and it may be that their understanding of this is improved as they develop. Use your knowledge not to give or tell answers, but to guide the class to where they should be.

9.
PROPERTIES
& USES OF
MATTER

Learning Objectives

Having read this chapter you should be able to:

- have a good understanding of the physical properties of matter
- explain why materials are chosen depending on their properties
- have a good knowledge of what dissolving means
- understand the differences between solutions and mixtures and how to separate these

Physical properties of matter

In discussion with the class, identifying how we can describe objects should lead to certain characteristics that are common or easily recognisable. These common characteristics are referred to as 'properties'. Different materials can have similar properties, and children will confuse simple materials because they have similar properties or characteristics.

The properties that the class may identify include the following:

Colour: Size: Shininess: Smoothness: Hardness: Mass: Weight: Solid: Liquid

Some of the terms here are not properties but describe the state of matter. To avoid confusion ensure the materials that you introduce the topic with are all solid. It is easier to use this as a stepping stone to further work.

If we consider the properties identified above and remove the states of matter we are left with:

Colour: Size: Shininess: Smoothness: Hardness: Mass: Weight

These 'properties' can be subdivided more into properties: those that depend upon the amount of matter that is present and those that do not.

'Mass' for example depends upon the amount of material but 'Colour' does not.

Intensive properties are those that DO NOT depend upon the amount of matter.

Extensive properties are those that DO depend upon the amount of matter present.

These terms are beyond the scope of schoolchildren, but are introduced here to assist you in familiarising students with these terms should you come across them in preparation and planning.

This section will focus more closely on some of the physical properties of matter. We introduced this in the previous chapter but only in order to establish a basis for the following content. For simplicity's sake we will take a physical property as that which can be observed or measured without dramatically changing the material in any way. There are many physical properties which we can use to classify and compare materials, some of which will be beyond the scope of primary school children. The key requirement in this section is to allow learners to categorise and classify materials in the world around them. By doing so they will gain experience of the range and variation of the world around them, appreciate the need for materials to have certain characteristics that we wish to utilise, and begin to develop knowledge of how we can adapt materials to our own needs. The general approach suggested here is one of allowing children to experiment and describe their ideas regarding the nature of materials, which will then allow them to develop their own concepts and understanding. The chapter outlines some of the key characteristics to consider when classifying materials, but the list is not exhaustive.

An obvious one here is hardness. Hardness however is not that simple as some materials are very difficult to scratch but are easily broken, such as glass.

Hardness can be described as a measure of an object's resistance to being scratched or damaged. A brick for example would be considered hard as it appears to be strong and hardwearing and able to withstand rain, wind, and changes in temperature. A simple and effective way of comparing hardness is to attempt to scratch one object with another.

Scratching a piece of metal such as a coin with a plastic straw would give us an indication of which is the harder material. The straw will not scratch the coin but the coin will scratch the straw. This allows us to make a comparative test of their hardness. This could be extended to comparing six or seven different everyday materials and noting in turn which materials can scratch the other. Possible materials for this could include card, a pebble, a coin, a piece of wood, a cloth, and a piece of plastic. Children can examine which materials are able to scratch others, and develop a small scale or table of results to indicate which materials may be grouped as 'hard' or which materials are harder than others.

An advantage to this is that there is no real list of materials based on their 'hardness'. The results that learners obtain are 'correct' and so they can develop their understanding from their own viewpoint and observed results.

Hardness is a very complex thing to determine. There are certain scales and tests which can be performed but their use is generally quite specialised. The Mohs hardness scale is used to make comparisons between certain rocks and minerals, based on experiments similar to the one described above. This can be used in the classroom when investigating rocks and stones. Can one type of rock scratch another? The scale is a simple comparative one. It indicates which material is harder, but not by how much.

Strength

Some materials can be described as strong. In school terms a definition around how easily a material can be bent or damaged or broken would be acceptable. There are associated difficulties here in that our general definition of the word 'strong' when referring to materials would generally include how much one could pull or stretch a material before it snaps or tears (destructive testing). Simple experiments can be done here by hanging a thread for example and then adding small weights to the bottom and testing the thread until it snaps. Destructive testing is always good as it interests the class and there is a definite observable conclusion to the experiment. A difficulty here is that the tensile (or stretching strength) of small threads can be great and it can take a relatively large load to actually break a thread. (That is why they are used to hold garments together. They stop your sleeves from splitting for example.) An interesting investigation is also to test the strengths of various supermarket bags to see how much these can carry. The difficulty with this however, is that they can withstand a very heavy load!!!

Different types of paper would be suitable for destructive weight-carrying testing. A hole can be made in the paper, near one end, and a string passed through to form a loop. The paper can be hung from a clamp or simply held, and small weights can be added to this loop until the paper rips. Newspaper, A4 paper, tissue paper etc. will all give reasonable and measurable results.

A good investigative activity here is to examine small strips of paper, card, wooden splints and plastic. If we obtain similar strips of these materials we can investigate how effective these are at supporting weights (between two 'blocks' for example) (see Figure 9.1).

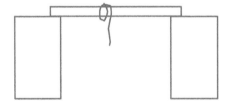

Figure 9.1

Small weights can be added to the thread and pupils can examine how much the paper or card bends, folds etc. The materials do not need to collapse, but pupils could be asked to describe how those materials behaved when they were being loaded, and also what happened to them when the load was removed (did they change back to their original shape?). There is no relationship to be determined here. It is a simple investigation into how materials bend or break.

A further investigation could be done using only a sheet of paper. In this case the paper should be folded into different shapes. Using sheets of 200mm × 100mm paper, allow learners to fold or twist the paper into a loop, a square, and a triangle. They can then investigate which shape can carry the greater load in a similar experiment to the earlier one. The key point here is that the shape of the 'beam' is important. There are certain shapes which are 'stronger' than others. The explanation of this is complex, but the key teaching point is that we can make things stronger by forming materials

Photo 9.1

iStock.com/Nattakit

into certain shapes. A steel I-beam is one of the strongest shapes for its weight and that is why we use these in construction.

Density

Density is the term given to how heavy an object is for a certain volume. It is a complex concept for young people. Many young children will confuse density and weight or occasionally see them as transferable. A large and relatively heavy football will float on water but a small screw will sink to the bottom. In technical terms the density of a material is determined by dividing the mass of the material by the volume. A block of stone for example may have a mass of 8.5kg and a volume of 0.006m³. This would give a density for that material of 8.5/0.006 = 1416kg/m³.

This may seem a high value but the density of water is generally 1000kg/m³. This material is slightly denser than water so it would sink if placed in water. This could be described or explained in terms of a stone being thrown into a puddle and it sinking to the bottom. The actual values for density are not important but the relative values (i.e. how does this material compare to that?) are worth debating.

Immersing objects in water and seeing if they sink or float is a simple way of determining whether a material is denser or less dense than water. It also allows discussion as to the reasons why objects sink or float and the children's descriptions or explanations will allow you to gauge their level of understanding quite easily. In general, metals and rocks are the densest of common materials and gold jewellery, lead weights and iron nails are good examples of dense materials.

The density of human beings is just slightly less than that of water. We can float but most of our body is submerged so that we need to tread water in order to keep our mouth and nose above the water line. Seawater has salt dissolved in it and this makes it denser than drinking water. As a result it is slightly easier to keep your mouth above the water line. The Dead Sea is so salty it is extremely dense in comparison to tap water. It is very easy to float in the Dead Sea.

Photo 9.2

iStock.com/RuslanDashinsky

People can lie back with enough of their body above the surface so they can read books and papers. Attempting to do this in a swimming pool would result in most of you being under the water! There are a number of simple activities which can be done regarding comparative densities.

Examples of density activities are shown on the companion website
https://study.sagepub.com/chambersandsouter

- An egg will sink in fresh water but when salt is added to the water its density increases. The egg is now less dense than the water and rises to the surface.
- Non-diet juice has about 9 to 10 teaspoons of sugar dissolved in it. This makes it slightly denser than the diet equivalent. If we place the diet and regular can in a bowl of water, the diet can will float and the regular can will sink due to the sugar dissolved within it.
- Raisins are denser than water and lemonade and will sink to the bottom. But after a while the raisins in the lemonade will start to rise and fall. Carbon dioxide in the lemonade will come out of solution and form on the raisins. Once enough bubbles

have appeared the raisins will rise to the surface. Once at the top the bubbles will burst into the air and the raisins will sink down and repeat the process.

- A density ladder is made of liquids that do not mix easily and have different densities. As a result they float on top of each other and the boundaries where they meet can be seen easily.

Solubility

Solubility is a measure of how much of a material can dissolve in another. The material that is dissolved is referred to as the *solute* and the material that it dissolves in is referred to as the *solvent*. Water is the most common solvent but there are others we use in everyday life, such as nail varnish remover or white spirit for removing paint. In primary school terms water is key and safe to use. When solids dissolve in water the particles of the solid 'separate' and disperse throughout the water. This is why sugar for example changes from a white crystalline solid and becomes transparent when dissolved in water. The sugar has not disappeared, it has separated and spread like a mixture. If we were to allow the water to evaporate the sugar would be seen again as it will come out of solution as the water reduces. Salt can also dissolve, as can many other chemicals, but the reasons as to why certain chemicals dissolve whilst others do not are complex. On a more microscopic view, when one substance dissolves the solute molecules are surrounded by the solvent. Once a chemical has dissolved in another chemical it has formed a *solution*. A solution is a particular type of *mixture* where the substances are physically mixed but not chemically, and has the same uniform appearance (i.e. it looks like a clear or coloured liquid, though not cloudy).

A mixture is a combination of two or more substances that are *not chemically* combined. They also do not need to be combined in certain proportions. You can have 'even' or balanced mixtures where both substances have roughly the same amount, or a mixture where there are only tiny amounts of one substance within large amounts of another. Most naturally occurring materials are mixtures, and if there is a particular material we wish to extract, we will have to separate those mixtures somehow. All mixtures can be physically separated into their individual constituent parts.

Homogeneous mixtures

The term homogeneous means 'the same' or 'similar'. An homogeneous mixture has the same uniform look or appearance to it. An example of this would be vinegar; it has ethanoic acid dissolved in water. It has the same appearance throughout. Many homogeneous mixtures are referred to as solutions. Diluting orange, soft drinks etc. are examples of these.

Heterogeneous mixtures

The term heterogeneous means 'different'. A heterogeneous mixture consists of substances which are notably or visually different. Peas in a pot of water, carbonated
water where you can see the bubbles etc. are heterogeneous.

Technical terms and names relating to types of solutions are not appropriate for use
with learners, but as an emerging teacher, exposure to a range of classifications and definitions will give you the confidence and subject knowledge to explore more if required.

Investigating solutions

There are a number of ways children can investigate and explore solutions. Water is
a safe and wide-ranging solvent. Gather together a range of materials (e.g. sugar, sand,
some dirt and salt).

Early experiments should focus on what does dissolve and what does not. This is
not that simple. You will have to establish an experimental procedure which the children can follow and carry out confidently. One such suggested procedure is to use
transparent plastic beakers and small splints as stirring rods. Work towards educating
the class to establish a fair and reasonable procedure – this is a recurring theme.

A suggested procedure is:

- Half fill beakers with cold water.
- Add a small spoonful of each substance to the beaker.
- Stir the water for a short time.
- Allow the water to settle for a few minutes.
- Check the base of the cup for any material still remaining.

Whilst not overly complex learners can be confused by the dirt being in solution.
Some dirt will appear to dissolve in the water, possibly making it cloudy, but when
the water is allowed to settle there will be small amounts of grit appearing. Some
children may suggest that whilst the water/dirt is being stirred it is a solution, but
when it settles they can discuss if any of the dirt has dissolved. Refer to washing their
hands and discuss the colour of the water after their hands are clean. Explore whether
the dirt has dissolved or if the water merely helped remove the dirt from contact with
their fingers and hands. Dirt does not really dissolve in water but water helps dislodge
the dirt from their skin. If we add soap or detergent it generally binds with the dirt
and the water can then help remove the dirt and it washes down the drain.

A teaching point here is that all the contents of the beakers are mixtures but the
ones containing salt and sugar are solutions.

This approach can now be expanded and other avenues explored. For example:

- Do salt and sugar dissolve by the same amount?
- Does more dissolve in a larger amount of water?

- Does the amount of stirring affect the dissolving?
- Does hot water dissolve better than cold water?

There are other examples or factors we can vary here, but the key skills are those related to scientific investigation and experimentation. Is it fair? Can the children manage the apparatus? Are their results recorded appropriately?

Dissolving and temperature

A simple experiment can be attempted here to compare how the temperature of the water affects the amount a substance can dissolve. An initial investigation could also be organised around hot and cold water and sand.

Set up a beaker of hot water (hot tap water would be sufficient here) and a beaker of cold water.

Add a spoonful of sand to both of these and stir for a while. The sand will not dissolve in hot water or cold water. This allows reinforcement/confirmation of materials that are soluble or insoluble.

This easily leads on to an investigation of hot and cold water with a soluble material.

The procedure can be varied in this case. A straightforward investigation can be focused around adding spoonfuls of sugar or salt until no more will dissolve, and comparing the results for the hot and cold solutions. Hot water will dissolve more solute than cold water.

The results or implications of this allow discussions of why we use hot water when washing dishes or showering. Is it to help dissolve certain substances? Is it to dislodge grease or oil? Is it a combination of both?

An added benefit of this is investigating what happens when the hot water cools. Consider that 10g of a solute has dissolved in hot water but only 4g dissolved in cold water. What will happen to the additional 6g of dissolved solute when the water cools down?

The additional 6g will come out of solution. As the water cools the solute will begin to appear at the bottom of the beaker. This further reinforces the idea that the solute does not 'vanish' that some children may initially hold.

Separating mixtures

The example described above of a dissolved substance being retrieved and 'coming out' of solution is an example of a solution being separated back into its original or constituent parts.

Mixtures and solutions can be separated in relatively simple ways as the original substances had not combined chemically. Had the substances combined chemically they would not be mixtures, they would be 'compounds', and the processes involved in separating compounds into their original substances are generally more difficult.

Separating solid mixtures

Investigate a mixture of sand and pebbles. Combine a small amount of sand and some small rocks in a plastic beaker. The key to separation here lies in the fact that both substances are solid, so there is no dissolving, and that the substances are separated by their size. It can be easily seen that the pebbles and sand are just spread amongst each other.

Learners could be asked for suggestions as to how best these could be separated. One method of separation is simply to pick the pebbles out of the mixture by hand or to use some sort of tweezers.

A further question could be proposed by asking for a quicker or more efficient method of separating them. This leads into filtering or the construction of a type of filter or material with small spaces that will allow the passage of one substance but not the other. A range of materials can be used as filters, including sieves, simple wire mesh, or even some form of woven fabric if the spacing is appropriate. If the solids do not differ in size then filtering is not really a possibility.

Separating solid and liquid mixtures

Separating solids and liquids which are not solutions can also be done using a form of filtration. Sand and water for example form a mixture that would be difficult to separate by extracting using tweezers for example. We could separate this mixture by simple filtration using an everyday sieve and a beaker underneath.

If we pour the mixture through the sieve, the water passes through and the sand is trapped by the sieve. Both substances can be gathered and retained. A key issue here is if we wish to keep the mixed materials. For example, if we wanted to keep the sand we could allow the water just to pour down the sink, and if we wanted to keep the water we would have to pour it into a larger container and possibly return the sand to the earth.

Settling

A less efficient but alternative method of separating solids and liquids (which have not dissolved) is simply to allow the beaker to stand and settle for a period of time. The sand or dirt will gather at the bottom of the container and can then be tilted gently and the water poured carefully into another beaker. This would leave the sand in the bottom of the beaker, and whilst it is still wet and possibly under a millimetre or two of water you will have separated the main substances of the mixture.

Dirty water will also become remarkably clear in a short period of time by just allowing it to settle.

In the same way as above we can tilt the glass and pour off the clean water.

Panning for gold

Panning for gold is a relatively simple process by which we separate heavy solids from a liquid. Gold-bearing rocks are generally dug out or extracted in large quantities from a mine. The rocks are ground into small, gravel-sized particles. This gravel is scooped onto a fairly shallow pan and water is added. The pan is gently agitated and the water mixture is swirled around. The water is allowed to spill over the lip of the pan.

This gentle tilting and rotating empties the pan of the less dense material and the more dense material remains at the bottom. The denser materials are generally heavy sand with gold particles or flakes. It is very difficult for this to be a profitable method given the sheer time it takes to separate the gravel, but it does eventually separate a mixture albeit rather inefficiently. A prospector would then spend time picking out the gold by hand.

This process generally produced very small amounts of gold, and in order to make their fortune prospectors would have to find more efficient methods or hopefully pick out a large nugget from within the rocks.

Photo 9.3

iStock.com/BanksPhotos

Separating dissolved solutions

Thus far we have considered methods for separating mixtures which have not dissolved. The materials can be seen relatively easily, and we can detect when they are separating (i.e. they are trapped in a filter or falling to the bottom of a container).

When a material has been dissolved to form a solution, filtering and settling will not work. This however has to be observed by the class. It is not good practice just to say to the class this will be the case. Allow learners to experiment and determine this for themselves. Dissolve some sugar or salt in a container, allow the solution to settle, and then pass it through a filter or a sieve. The class should observe that the solution has not separated and has remained a solution.

The only way to separate this is to allow the solvent (water) to evaporate. This can be done by heating and allowing the solution to boil until the solute is left. This is not easily done in a primary school without access to Bunsen burners for example.

An alternative here would be to pour the solution onto flat trays or glasses. (Foil baking cups for cakes are good.)

Place these on or above a radiator or window ledge. Allow the liquid to evaporate and the solute can be easily seen.

This method of separation by evaporation is commonly used to harness seasalt from saltwater.

Salt pans or shallow lagoons are built near the coast in places where there are generally hot temperatures. They are relatively large flat areas with small earth walls.

Photo 9.4

IStock.com/manx_in_the_world

Seawater is allowed into the pans and is then trapped there.

Sunlight and wind evaporate the water, leaving the salt behind. Workers scrape the salt from the bottom of the pans and gather the salt in small mounds.

The salt is then taken away, packaged, and sold in markets and supermarkets. In these circumstances the water is allowed to evaporate, as it is the salt that is being extracted.

In other situations it is desirable to extract 'fresh' water from the salt water.

In warm countries such as in the Middle East for example, there is little rainfall and therefore less supplies of fresh water. In areas such as these there are large-scale 'desalination' plants which will extract fresh water from salt water. The simplest way is to distil the latter by boiling the water until it evaporates and then allowing the steam to cool and condense. The salt is a byproduct in this case. This requires large amounts of energy and as a result is very expensive. Another method involves using a form of filtration called reverse osmosis, but this is also expensive and requires the water to be subject to great pressures.

Summary

This chapter highlighted the main ideas when teaching the topic of matter and its properties to young people. It introduced some very difficult terms such as density and solubility. These terms are complex, and the concepts may be abstract, and for young people will possibly be beyond their understanding. This is acceptable. When introducing density for example, concrete examples of sinking and floating are used and alternative terms or words (such as weight or heaviness) from the pupils can be used; once these have become established and familiar we can extend their experiences. This section also introduced some of the properties of water which are explored further in later chapters. You may feel that some of the experiments are difficult or not appropriate within your own school experience. The intention has been that you have garnered the knowledge and confidence to explore this if the opportunity presents itself.

REFLECTION POINTS

- How could you introduce the concept of density to an upper primary class?

  - Density could be referred to as the weight of an object for a specific volume. There are a number of approaches you could take to address this. Take some samples of materials which could be placed in water and observe those that sink and those that float. The second phase would be to select materials that could be cut or moulded to the same shape (e.g. a cube). Take these cubes (e.g. of bread, plasticene, wood, plastic packaging, an apple) and have them all roughly the same volume (2cm x 2cm x 2cm). Ask the children to place these in order of the heaviest to the lightest. This introduces the concept of density in that the heaviest cube is the densest.

- How could you explain the difference between hardness and strength?

  - These two terms are interchangeable, but the key is to consider scratching/scraping/softness and bending/breaking separately. A strand of hair for

(Continued)

(Continued)

example is soft and can be folded and twisted. However it is very strong if you try to pull it. It has a good tensile strength. Glass is hard and not easy to scratch (you can rub your nail against it) but it can break easily. It is hard but not that strong. Steel is both hard and strong. Tinfoil is soft and can be torn easily. A number of materials are more difficult to categorise, but the discussion around the properties is what you wish to develop.

- What are the main physical properties we desire for the following items?

  - *Shoes* Hardwearing: waterproof: grip
  - *Jewellery* Shiny: does not react with skin or moisture: can be shaped
  - *Jacket* Waterproof: good insulator: flexible
  - *Spade* Hard: strong: does not corrode
  - *Frying pan* Good conductor: high melting point: non-stick

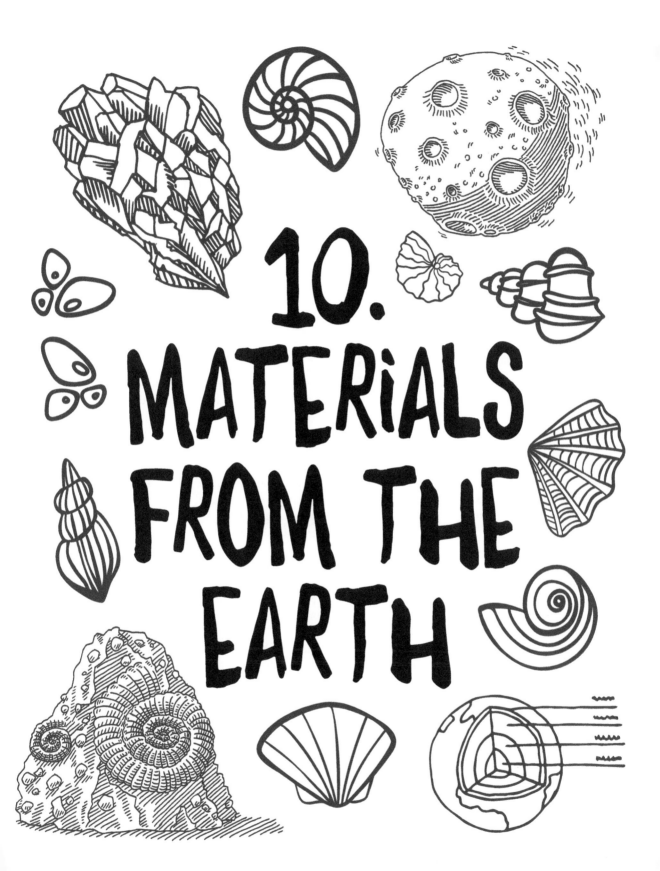

10.
MATERIALS FROM THE EARTH

Learning Objectives

Having read this chapter you should be able to:

- compare and group together rocks on the basis of their appearance and properties
- describe in simple terms how fossils are formed
- explain how soils are made from rocks and organic matter
- know the structure and composition of the Earth
- explain in simple terms how igneous, sedimentary, and metamorphic rocks are formed

Overview

Formation and composition of the Earth

The Earth was formed around 4.5 billion (4,500,000,000) years ago. It was formed from the dust surrounding our star which gradually came together through gravitational attraction. The relatively accurate age determination comes from the age of some minerals found in the planet which are believed to be amongst the first substances to form our solar system.

Small particles drew together to form larger and more attractive masses. This attracted more material and the more dense materials bound together at the centre, forming a rocky core, while less dense material formed near the surface creating a crust. During this phase the Earth was struck by a large body which scattered a lot of material in the crust and the layers below (the mantle) into space. This material gathered itself together to form the Moon.

When forming initially, Earth was subject to collisions by meteorites and internal volcanism. This explosion of materials and gases produced the beginning of our very early atmosphere. There was a lot of liquid water at this time and this condensed in the atmosphere, cooling the exterior of the planet, and forming the crust and the oceans. Around 3.8 billion years ago areas of the Earth's surface cooled and some extremely basic life forms may have started to develop. Earth's core was and remains incredibly hot, and the movement of molten rock beneath the surface led to the formation and separation of our continents. Even now our continents are still moving relative to each other, but thankfully on a very slow scale.

Structure of the Earth today

The Crust This is the hard outer layer where we live. The large land masses and oceans are all part of the crust. Its thickness varies but the continental crust which we live on is generally 20 to 30 miles thick (tiny in comparison to the diameter of

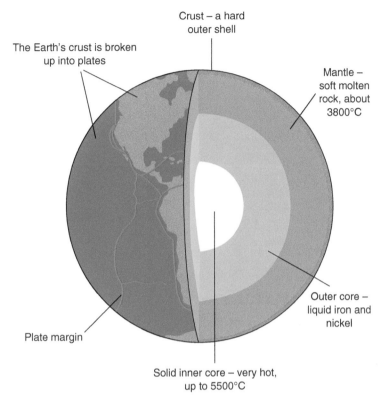

Crust – a hard
outer shell

The Earth's crust is broken
up into plates

Mantle –
soft molten
rock, about
3800°C

Outer core –
liquid iron and
nickel

Plate margin

Solid inner core – very hot,
up to 5500°C

Figure 10.1

the Earth) and the crust under the oceans is much thinner (~4 to 5 miles thick). The materials making up the crust are less dense mainly and it 'floats' on the layer below. All the rocks and minerals that we use for living are mined from the crust. The deepest ever borehole attempted made it to a depth of 7 or 8 miles, however most caves or mines are nowhere near that depth.

The Mantle This layer is beneath the crust and is less than 2000 miles thick. It is at a high temperature and the rocks, whilst not a soft, gooey liquid, do flow around the mantle. Hot, molten material flows upwards, and where the crust is thinner or has been fractured it gathers and cools to form igneous rock, or it can erupt as a volcano.

Outer Core The outer core is composed of molten metal and is about 1400 miles thick. Its very high temperatures result in the movement of molten iron and nickel.

Inner Core A solid ball with a diameter of approximately 750 miles. Probably an alloy of iron and nickel at an extremely high temperature and pressure.

An obvious question here is, 'Why is the core of the Earth still so hot?', especially given that the planet is 4,500,000,000 years old. Surely it should have cooled in this time?

This is a difficult one to answer or explain clearly. There are three main processes. The main one is the heat generated by the gravitational 'squeezing' of the planet when it formed has not been lost. This was a complex process but essentially the matter towards the centre of the planet was compressed so much it raised its temperature. It takes a very long time for the heat to transfer through all the layers and material that make up the Earth's core, mantle and crust.

There is another factor which is due to the frictional effects of denser material sinking and passing through towards the centre of the Earth. This friction generates huge amounts of heat.

The third factor is that there are radioactive materials deep within the Earth which release energy as they decay and this supplies heat to the core.

It is difficult to obtain very accurate information regarding what lies so deep underground, but the overwhelming message is that the Earth has a massive amount of energy under the surface, that this energy is pretty well insulated by the surface, and that it will not cool down anytime soon.

Structure of the surface

As indicated earlier we live on the crust or the surface of the planet. We live on continents and these continents are continually moving, albeit very slowly, and there are certain areas where they press against each other. These are generally the areas where earthquakes occur and they are a result of a build up of pressure from deep underground. Surface rocks 'break' under the stress of this. It has been suggested that the land that currently makes up our islands and continents was once all connected, and that the movement of these continents over millions of years has led to the shape of our land surface being what it is. Many mountain ranges were formed by the coming together of these areas, and the movement forced the surface upwards to form high altitude regions such as the Himalayas. Britain is not close to any area that is moving relative to another area, and as a result does not have any major incidence of earthquakes, and those that do occur are very small in relative terms.

The ground under our feet is a result of the movement of these plates and a number of environmental factors. The rocks and various types of soils which make up our environment are created by a number of processes, and can be broadly categorised due to the processes involved in their formation.

Formation of rocks

There are three main types of rocks: igneous, sedimentary, and metamorphic. All rocks can be classified as belonging to these three main types. Within these broad headings are a large amount of subheadings, but the text will focus on these primarily.

Igneous rocks are formed when molten rocks beneath the Earth's surface reach an area where it is cool enough for them to solidify. Molten rock beneath the surface is referred to as magma, and when this breaks through and onto the surface it is referred to as lava.

There many types of rocks formed this way, but examples of igneous rocks include granite, obsidian, basalt and pumice.

Photo 10.1 *Pumice stone*

iStock.com/AquaColor

In general, igneous rocks have many crystals which can be seen quite easily. Close examination of these rocks provides an opportunity for children to draw, describe and note any unusual or identifying features.

Sedimentary rocks

Sedimentary rocks are formed by the action of layers of materials which are on the Earth's surface. These materials are eroded or blown or washed into areas where they can collect. One layer may be settled down and over the course of time more layers are gradually deposited in that area. These 'sediments' are laid one on top of the other and the gradual pressure of these layers, over time, compresses and transforms the layers into rocks.

Sandstone, limestone, chalk, coal and shale are examples of sedimentary rocks. Sedimentary rocks can have different appearances but closely squashed, gritty materials tend to be sedimentary. Additionally, layers or strata are indicative of sedimentary rocks.

Metamorphic rocks

Metamorphic rocks are those which have changed in form somehow. Originally metamorphic rocks would have been igneous or sedimentary. Magma beneath the surface

Photo 10.2 *Sandstone*

iStock.com/Don Nichols

may have broken through between a layer of rock, and the intense heat and pressure from the magma altered the composition of the rock and made it change or 'metamorphose'. Additionally, two different rocks may be compressed together under great pressure to form a different type of rock.

Marble, slate and gneiss are examples of metamorphic rocks.

Photo 10.3 *Marble*

iStock.com/Sabrina Pintus

Soil

Whilst not a rock, soil is possibly the most common material that children will answer with if you ask them to discuss what is in the ground. Soil is a major component of our natural environment and has a number of constituent parts. Essentially soil is a mixture of rock and decaying plant material. Once rocks have been formed, environmental factors gradually break or grind down the rock into smaller and smaller particles. These small particles are then combined with decaying plant and other organic material and are affected by wind, sun, rain etc. All these processes combine to produce the soil we see in our gardens. Different climates and initial rocks combine to form different soils, and this is what creates the range of soil types worldwide.

Fossils

Fossils are the remains of dead plants and/or animals that have been trapped in a particular environment. Living material that dies generally decays and the matter is gradually eroded, decomposed and spread. However, under the correct conditions the dead matter may perhaps be trapped in sand or mud. Over time more sediment covers the remains and presses down upon them. Dissolved minerals within the rocks leech into the remains of the bones. The bones then change to mineralised fossils over millions of years.

Ultimately erosion or some other event uncovers the rocks and the fossils are discovered. Many driveways and flagstones are made from sandstone and limestone which are sedimentary. Often these stones will show the clear imprint of a fern or simple plant and these are described as fossils. Sadly, they are not. They are a sort of chemical stain left behind by the plant material. When the stone is exposed to the surface, the stain reacts with the atmosphere to reproduce the shape of the fern or plant when it was trapped in the sand.

ACTIVITY

Fossil making and fossil finding

This area lends itself to some fun and investigative practical activities which can be used in combination to provide a series of lessons around the nature of fossils and sedimentary rocks.

Making a fossil
You will need:

- A plastic cup or small plastic container
- Some modelling clay like plasticene

(Continued)

(Continued)

- Plaster of Paris
- A small plastic animal or plant (to be the fossil)

Place the clay in the bottom of the plastic cup and flatten it so it covers the base to a depth of 1 or 2 centimetres.

Press your small plastic animal into the clay at the bottom so it makes a mould or leaves a shape in the clay.

Mix a little Plaster of Paris with some water, then stir for a minute or so until it is smooth and slightly runny and can be poured. Pour it into your mould or shape in the clay. (You can pour it so that it just fills the shape you have made (1), or you can pour it so that it completely covers the clay (2).)

Leave this overnight. Remove the plaster from the clay, scrape off any loose sections, and you will have a fossil. If you followed step (1), the fossil is the actual shape of the mould and you can look at it and turn it over and such. This can be used in another activity.

If you have followed step (2), the fossil will have its own little base. You can use this base as a small stand and it can be stored as a fossil exhibit. The fossil can be painted or coloured to give an indication of what it was like when alive, and can be used in a mini fossil grove or prehistoric valley.

Fossil excavation

This activity involves burying the already made fossil in a range of soils and sands and trying to extract the fossil without breaking it.

You will need some cornstarch which forms the basis of the material the fossil is encased in.

Cornstarch mixed with water (2:1) can produce a soft, semi-solid material that can be used to bury the fossil. It is a thick, gooey material but will go a little hard. It can then be placed in a tray, possibly under sand for example, and the class can then try and extract the fossil using small spoons, brushes and so on. This gives an indication of what palaeontologists and fossil hunters have to do to extract old items safely.

The planet's natural resources

In order to discuss how we use and misuse our natural resources, it is worthwhile to consider what is meant by a natural resource. A simple explanation would be that natural resources are materials produced by the Earth which we can use to make more complex man-made products.

Table 10.1

Resource	Product
Air	Wind energy, combustion
Animals	Foods, clothing
Plants	Foods (fruit/veg), timber
Fossil fuels	Heating, energy generation
Water	Drinking, cleaning, power generation
Minerals	Metals, jewellery
Sunlight	Solar power, photosynthesis
Rocks	Building materials

By definition natural resources are materials which we do not manufacture and that occur in nature. We use these resources in order to make our life more comfortable or our existence easier. There are many natural resources but this is a list of the most common:

We use our resources to improve our general living conditions. Some resources can be used but will be replaced or renewed naturally. These resources are referred to as *renewable* or *sustainable*. Air, water, timber, animals and so on will continue to be renewed for the most part. Essentially these resources are being replaced naturally at a rate equal to, or greater than, the rate at which we use them. The depletion of these resources is generally not a major concern. There are occasions such as droughts or famine for example where there is not enough food or water to sustain a population, but this is generally localised. It happens in a certain, specific area and there is perhaps enough food elsewhere, but there are difficulties in getting that food to where it is needed. Very often the problem is caused by local management of the resource being affected in some way (e.g. civil unrest).

Fossil fuels

How fossil fuels were formed

Fossil fuels, contrary to what some children may believe, are not the remains of dinosaurs that died in a swamp and were then crushed like a sedimentary rock. Generally most of the fossil fuels we find today were formed millions of years before the first dinosaurs.

Fossil fuels, however, were once alive!

They were formed from the remains of plants and animals that lived hundreds of millions of years ago. The world was somewhat different then. The continents we live on and the seas we travel on were forming and changing (see the section on the structure of the Earth). The climate was warmer and wetter, with huge areas of water, trees and vegetation. There were some unusual animals on land and lots of tiny organisms called proto-plankton in the seas.

When these tiny organisms died, they sank to the bottom of the waters they were in, and became buried under layers and layers of mud, rock and sand. They became trapped in these layers and therfore 'sedimentary'. Eventually, hundreds of metres of layers of rock and soil covered them. In some areas, the decomposing materials were pressurised under the seas.

Over millions of years the dead plants and animals slowly decomposed into new materials which we refer to as fossil fuels. Different types of fossil fuels were formed depending on which combination of animal and plant material was present, how long it was buried, and what temperature and pressure they were exposed to.

Oil and gas were created from organisms that lived in the water and were buried under ocean or river sediments. Over time, heat and pressure from the weight of the layers above (and some bacterial action) changed the material. Initially a thick liquid (i.e. oil) was formed, but in deeper, warmer layers, the material was changed into what we refer to as natural gas.

Whilst these fuels were formed deep underground some made their way closer to the surface. Pressure and the movement of rocks underground eventually forced the gas and liquid oil upwards. This continued until they reached the surface, or they rose but were trapped by a solid dense layer of rock which stopped their progress. Almost all of our oil and gas is found under these dense rock layers.

Coal was formed by a similar process but initially produced from the remains of dead trees and plants, not animal material. This process has been under way from around 300–350 million years ago. However some coal began the 'process' more recently, and this gave rise to different types of coal being formed.

Anthracite is the most energy-concentrated type of coal. It is shiny and difficult to ignite, but burns with a small blue flame and produces a large amount of energy per kilogram. It has the fewest impurities and is good for power generation and use in homes. It accounts for about 1% of the world's coal reserves and has been mined extensively in the UK.

Bituminous coal is a lower-grade coal and does not produce as much energy per kilogram as anthracite. It is softer and less shiny than anthracite and contains some impurities including a tar-like substance called bitumen. It is generally treated to remove impurities and is then known as *coke*. This is used in furnaces in the iron and metal industries.

Lignite is a lower-grade coal which is brown in appearance, softer, and produces less heat per kilogram than the other types. It is used mainly in power stations for electricity generation.

Peat is not coal but has undergone a similar process in that it is the decaying organic matter that over time has formed a type of fuel. The process takes thousands rather than millions of years, and has not undergone the pressure and temperature 'conditions' that coal has. It is used as a source of fuel in many areas. It is not really a renewable source as it is extracted more rapidly than it forms, but some bodies regard it as a 'slow-renewable fuel'. Peat has been used throughout Britain since the Bronze and Iron Age. Its use in Scotland and Ireland has a long history, especially in the remoter areas and islands where it was stacked and dried and used for heating and cooking.

Effects of fossil fuel utilisation

When oil, coal, gas or peat is burned energy is released. In simple terms a fuel burns by reacting with oxygen in the air to produce heat. It also produces water vapour and carbon dioxide (CO_2). Whilst not toxic or poisonous, carbon dioxide has been identified as a major factor in the increase in the average temperature of the planet. Carbon dioxide is a gas which absorbs radiation from the Sun and re-radiates this in the lower atmosphere. This is what is referred to as the greenhouse effect and the labelling of carbon dioxide as a 'greenhouse gas'. The increase in the percentage of carbon dioxide in the atmosphere has been directly linked to the industrialisation of the planet over the last two to three hundred years. This industrialisation has led to an increase in the average surface temperature of the planet by about 1°C. The ten warmest years (up to 2014) are shown in Table 10.2.

Table 10.2

Warmest in order Records (1880 – 2014)	Year
1	2014
2=	2010
2=	2005
4	1998
5=	2013
5=	2003
7	2002
8	2006
9=	2009
9=	2009

Source: www.ncdc.noaa.gov/sotc/global/201413

Interestingly, all of these 'warmest' years have occurred from 1998 onwards. Overall, the global annual temperature has increased at an average rate of 0.06°C per decade since 1880, and at an average rate of 0.16°C per decade since 1970. Much of this warmth is due to increased global ocean temperatures. Land temperatures have also increased, and in general the increase in land temperature has been greater than those for ocean temperatures.

This 1°C may not appear much, but the impact of this additional energy may lead to arctic melting, increased moisture, more frequent and more powerful 'extreme' weather events and the like. The effects of increased temperature throughout the world are varied. In general terms there will be greater warming over land, mainly in the Northern latitudes. This will lead to increased rainfall in Northern areas, with probable associated flooding and extreme monsoons and droughts in other areas. It is predicted that droughts will become more frequent in areas of Europe, the Middle East, Central America, Asia and Australia. Increased temperature will impact upon farming, with plants being harvested earlier than 'normal'. This has already happened with wheat and maize production in some countries. Rising water levels may also make many areas uninhabitable, such as low lying areas in Bangladesh or small island states such as the Maldives. Southern Europe may become so hot that some crops will not thrive, but Northern Europe may be able to have more diverse crops as it too warms. As this is such a slow process the rise in temperature will be difficult to curtail. Even if we were to reduce carbon emissions to almost zero the process is in place and would take years to halt. A leading question in this area is how high will the increase go? Again, this is almost impossible to predict.

It is very difficult to directly attribute a direct 'cause and effect' to carbon dioxide and temperature rise, as the actual mechanism is very complex and the computer modelling contains many variables, but most scientists find the evidence for this overwhelming. There are some 'climate deniers' however, and they believe we should continue to use the resources for as long as we can.

The *United Nations Framework Convention on Climate Change* (UNFCCC) was designed to promote and encourage practices which will help to prevent dangerous climate change, and countries throughout the world have adopted a number of policies designed to reduce greenhouse gas emissions and assist in the adaptation to global warming.

To reduce the amount of fossil fuels we burn, we can reduce our energy requirements and generate more electrical energy by renewable methods. Insulating homes, ensuring new buildings are constructed efficiently in terms of their resource use, and continually seeking ways to reduce our use of fuels will lower the 'carbon footprint' we cause in our everyday lives. Many items now are referred to as carbon neutral or eco-friendly in this respect. Current measures taken by countries will not have any impact on the average temperature in the short term but the hope is to slow down and halt the rise in temperature in the medium term.

Energy conservation and sustainability

How long will our fossil fuels last? This question is impossible to answer with any great certainty as there are a number of factors which can increase or decrease the length of reserves available. These include the following.

Factors reducing the time until we reach shortage

- Economic growth
- Industrialisation of third world countries
- Unrest in oil-producing areas

Factors increasing the time until we reach shortage

- Energy conservation approaches in industrialised countries
- Growth in nuclear power
- Growth in renewable sources technology
- Increased exploration for oil/gas reserves

Searching for estimates can be time consuming and difficult. It is a political area and many sites are making points from a particular agenda, and the volatility in various factors can have a complex effect. For example, the fall in the oil price from approximately $100 per barrel between 2011–2014, to $30 at the beginning of 2016, resulted in a drop in petrol prices. This encourages people to drive more as it is cheaper, possibly increasing demand. This drop in the price of oil makes it uneconomical for firms operating in certain areas, as it costs more to extract crude oil than they get paid for it. This also slashes oil exploration as there is no point looking for oil if it will cost firms money to provide it. This will reduce production until the price makes it profitable. However, the fall in the oil price has lasted for over a year as other countries increase their production to retain income. If the price were to increase dramatically so would exploration and production, and therefore attempting to determine how long resources will last is very problematic.

Summary

This chapter examined the formation of the planet Earth and a number of the resources that are prevalent in the crust. The formation of the main rocks that exist and how some of the more important resources formed were described in some detail to allow teachers to emphasise our dependence upon these raw materials. Our energy needs and usage were also discussed to highlight how and why we can affect our own planet, and what our options are in order to minimise any future negative impact.

REFLECTION POINTS ·······································

- Describe in simple terms the three main types of rock.

  - Igneous: Formed from molten magma which cooled and solidified. Either underground or on the surface.
  - Sedimentary: A layered or compressed rock formed from other rocks and materials which have been deposited in a certain area. Over the course of time more deposits are added and the pressure forms a new material.
  - Metamorphic: Existing rocks are exposed to extreme temperatures and pressures and have their structure changed dramatically as a result.

- How would you explain to a pupil that oil is a non-renewable resource?

  - The problem is that it takes millions of years for oil to form. We are using it at a rate that is unsustainable. We are using more than is being renewed. No new oil has been created since we started using oil approximately one hundred years ago.

- What could we as citizens do to ensure we can cope with our future energy demands?

  - Minimise our own use. This saves energy but also helps us financially by having smaller energy bills.
  - Recycling is good. It requires less energy to renew plastic bottles for example than it takes to prepare new bottles.
  - Insulating our homes is good. It reduces heat loss and keeps our homes warm. Some loft insulation is made from recycling other materials.
  - Modern vehicles are now very efficient: they have battery power to reduce emissions and can travel great distances on smaller amounts of fuel.

11.
CHEMICAL
CHANGES

Learning Objectives

Having read this chapter you should be able to:

- describe what a chemical reaction is and how you know one has occurred
- have a good understanding of what an element and a compound are
- identify some common instances where chemical reactions have taken place
- perform a range of simple experiments showing a chemical reaction has taken place

Chemical reactions

What is a chemical reaction and how do we know that one has taken place?

When a chemical reaction occurs one set of substances has changed to another. Examples of this are:

- materials burning in air
- potassium reacting with water
- iron rusting
- frying an egg

Almost everything we see involves some sort of chemical reaction. The vast majority of the materials we rely upon in order to live and work are the result of a chemical reaction. The concrete in our bridges, the steel and other metals we use in cooking for example, the plastics used in packaging etc. are all the result of chemical reactions.

In simple terms however we need to examine the basics of this and look to identify how we can tell if a chemical reaction has taken place. In general terms we must look to see evidence of a change of some sort.

- New chemicals may have a different appearance – rust does not look like iron.
- If we have combined liquids, there may be a solid appearing (precipitate).
- Bubbles may be given off.
- Energy is released (or occasionally taken in).

These are examples where a chemical change has taken place. This is different from a physical change. For example, if you tear up pieces of paper you will still have the paper, albeit in bits, but if you *burn* the paper you will be left with a pile of ashes and you will not get back to the original material. Children will often confuse melting and dissolving (e.g. as examples of chemical change where they are in fact examples of physical change).

This topic can also offer a wide-ranging scientific experience for children in that there are 'experiments' which the class can take part in which will allow them to fully explore the range of skills and techniques that as teachers we wish to encourage. Whilst there are many activities which are 'experiments', it is the association of doing things with chemicals and burning etc. which is closely linked to the idea of scientists.

Background knowledge

All materials are composed of atoms and molecules, and they can exist as simple substances with perhaps only one constituent through to hugely complex structures with many different substances.

The simplest materials are *elements*. An element is composed of only one type of atom. Oxygen for example contains only oxygen atoms whereas carbon dioxide contains carbon and oxygen atoms. Carbon dioxide is a *compound*. This means the two elements, carbon and oxygen, have combined to form a new chemical. There are currently only around 118 elements, and all of the substances we use, breathe, eat and emit are composed of some of those elements. In reality though far fewer than the 118 elements make up the materials we are familiar with.

In a primary school context, children can be exposed to the materials that make up our world and how science has developed to use those elements for our everyday existence. All elements are pure in that they cannot be separated into other chemicals. However not all elements are found in their pure state. Most will have combined with other elements. Scientists have spent many years trying to identify and extract pure elements from their compounds.

The periodic table groups all the known elements according to their properties and atomic structure. This allows activities related to classification and discussion around the differences and similarities of chemicals. Chemical symbols are used as a shorthand version of writing the chemical name, and this allows investigative activities where the class can hunt for the name of the chemical given the symbol and vice versa.

For example, asking children to investigate certain chemicals and describe these and their uses will allow them to become familiar with the language of science and scientific enquiry. Young children's knowledge of elements may be limited and investigations should be chosen carefully. For example, in a project on metals or jewellery they could gather information on what makes certain metals wanted or attractive. They would then notice that these are closely grouped on the periodic table and that other less well-known metals (such as palladium for example) are also used in jewellery. The light gases, hydrogen and helium, are at the top of the table and this is due to the small mass of the atoms of these gases.

Many websites contain interactive periodic tables showing the history of discovery, perhaps along with who discovered the element and why it received its name.

Lead, for example, has the symbol Pb as it was called plumbum by the Romans. A *plumb bob* is a tool which gives a vertical line. It is a piece of string with a lead weight on the end and is hung from a wall. The string forms a perfectly straight vertical line that builders use to check buildings.

The word copper comes from *Cuprum* which was the Latin name for Cyprus. The Romans mined much of their copper there. Copper has the symbol Cu.

A less well-known element, strontium (Sr), was named after a tiny village in Scotland. A mineral was mined from the area around Strontian and this metal was extracted from it.

Alloys

There are also mixtures such as alloys which are not compounds but have materials fused within them that give them certain properties. Steel for example is an *alloy*. We take iron and heat it until it melts and then add carbon or nickel or chromium, for example to make a much tougher material which can be used in structural components in buildings or machines. Our cutlery for example is mainly stainless steel, which means it is hard and strong enough to cut through food but can also withstand hot temperatures and the chemicals used in dishwashers without changing its colour or corroding. Gold jewellery is also an alloy. Nine-carat gold contains more silver and copper than gold and these other metals, which are also elements, help make the gold more hardwearing and can also change its appearance into the colour we want.

As indicated earlier, when elements or compounds combine with other elements or compounds a chemical reaction takes place. New chemicals are then formed.

A very common chemical reaction is burning or combustion. In burning, a flammable material such as wood or a candle combines with oxygen in the air. This reaction gives off heat and light and produces new chemicals. Fuels such as oil or coal will burn for some time and generally 'change' into carbon dioxide and water. It may seem odd that burning fuels produce water. This does occur but the water is generally in vapour form. When our gas central heating boilers operate you can detect the water vapour from the outlet from the boiler. On colder days this steam/water vapour can be seen from the outlet pipe.

If you place your hand near to the outlet you can feel the warm air pass over your hand, but if you examine your hand you will notice it is slightly damp as the water vapour condenses on your palm. This production of carbon dioxide and water occurs when there is a clean flow of air (oxygen) and all the fuel is fully burned.

However, in many circumstances parts of the fuel may not receive a good flow of air and then only partially combust. In this situation, soot and carbon monoxide, along with carbon dioxide, are produced. Carbon monoxide (CO) is poisonous. If we inhale it, it is absorbed in our lungs and combines with our red blood cells, thereby reducing our blood's ability to carry oxygen. We will then feel dizzy, sick, tired and/or confused, and ultimately become unconscious and possibly die. Carbon monoxide has no smell and is difficult to detect. Thankfully CO poisoning is not common, but in older houses with boilers or fires that are not well maintained it can still occur. If you notice black, sooty marks on the front covers of gas fires for example, soot marks on the walls around boilers or fires, or yellow instead of blue flames from gas fires, then there could be incomplete combustion happening.

The terms 'burning' and 'combustion' are often taken as identical but there are some differences in detail. For example, a piece of coal burns in the fire releasing light and heat whereas a drop of petrol in an engine undergoes combustion. Both require oxygen and release energy, but in a car's engine it is more efficient and there are no flames.

You can demonstrate simple burning experiments with a class or possibly allow them to do the experiments if you feel confident enough.

You can demonstrate what happens when burning occurs quite simply.

ACTIVITY

Burning

You will need a small tealight candle and a couple of glass jars (coffee jars possibly) of different sizes. Light the candle and place the jar over the flame and ask the class to observe what happens. After a few seconds the flame will dim and go out. The burning fuel of the candle has combined with the oxygen and burned until all the oxygen has been used. Once the oxygen has been used and no more burning can take place, the candle extinguishes. The 'air' inside the jar has little or no oxygen and if you quickly place the 'used' jar over another candle it will also go out. No oxygen, no burning, no chemical reaction. This can be repeated with jars of different sizes to investigate how long it takes for the candle to extinguish. The larger the jar, the greater the time the candle will stay lit. This may seem obvious as there is more oxygen in a larger jar so the burning will continue for longer.

You can also do a simple experiment using a jar and a candle where you have a small saucer with some coloured water (orange juice would do). Pour a small amount of orange juice into the saucer and place the candle in the juice. Light the candle then place a jar over the candle. The candle will extinguish after a few seconds and then the orange juice will slowly be drawn into the jar.

(Continued)

(Continued)

The explanation

The candle reacts with the oxygen in the jar and burns, heating the air inside the jar. This expands and squeezes its way out of the jar (you may observe little bubbles appearing in the orange juice). Once the candle has gone out, the air inside cools down and contracts. This reduces the pressure and in turn draws in some of the orange juice in the saucer. This can be an eye-catching experiment!

In these cases we have investigated how the burning candle has changed and also how the air itself has changed due to the chemical reactions taking place.

All fuels do this whilst burning or combusting. Oxygen reacts with the fuel and releases energy as CO_2 and water vapour.

Respiration

We can take this a stage further by investigating which foods have more or less energy stored within them. To do this we simply need to take a piece of food and hold it in a candle flame until it starts to burn, and then observe how brightly or for how long it burns. Simple foods like bread, crisps, sugar and sweets are good for this as they burn relatively easily. You could also experiment with dried carrot or fruit. Fruit and vegetables do not burn well, and this can be related to our energy needs and how our diet affects our growth and development.

This should show that some foods 'contain' more energy than others and highlight foods that can supply us with lots of energy. The logical continuation of this is that we also 'burn' foods but in a different way. You could emphasise the fact that when we breathe in we take in oxygen and when we breathe out we exhale CO_2 and water vapour which is similar to burning and combustion. The process by which we obtain our energy needs from food is called *respiration* and in introducing this idea to students you can make comparisons to fuels releasing energy and the like. There is obviously no flame when we respire; it is a much more gradual and slow process and the energy is not released directly but via a number of mechanisms. It is more of a biological process than a chemical process, but it is worthwhile making the comparison with your classes. An interesting experiment here is to determine if we emit water vapour. Ask the class to breathe onto a cold window pane. 'Condensation' will be formed on the window. You could also ask the children to breathe into a polythene sandwich bag a few times. The bag will soon cloud over on the inside and moisture will be seen. This is due to chemical changes taking place in our bodies.

Reactions with water

Most metals do not react easily with water and this can be investigated by placing a range of metals in small beakers or containers of water and observing them over the course of a few days. Tacks, paper clips, coins, iron nails etc. can be left in the water for some time without any change appearing to take place. Changes to the surface of most metals are a slow process and these experiments would support this. Iron however will react fairly quickly where there is some moisture and some air present. Placing iron nails on a damp paper towel for example will result in some reddish brown rust appearing in a day or so. Classes can describe the appearance of the nails as these gradually react over the course of a week.

Iron fencing and railings are covered with a thick layer or layers of paint to protect them from rusting. Over time however air and moisture can penetrate the paint and react with the underlying iron. When water reacts with iron an oxide of iron is formed and these molecules are larger than molecules of iron. As a result they expand and flake away from the original iron, exposing more fresh iron in the process. This leads to what look like bubbles in the iron. This is a physical sign of rust under the paint. An interesting activity is to get the children to examine iron and steel handrails and fencing throughout the school, especially outside. Observe where the rust is formed, then note the age or condition of the metal and if the paint or outer layer has flaked off. This is an opportunity for observing and recording the condition of metal objects around the school. If there is a badly rusted pole there are opportunities to draw, describe and possibly predict what will happen if the rust is untreated. The term 'rusting' really only applies to iron and its alloys such as steel. Other materials do react with oxygen and water but this is referred to as a more general 'corrosion'.

Copper for example does not rust, but over a number of years the exposed copper will turn a greenish colour, which is the copper reacting with oxygen and other carbon compounds from the air. It is called verdigris and actually helps protect the copper underneath the layer. This is the reason copper (alloy) coins tarnish or appear to go dull after a while.

Rust is brittle and fragile, and as more of the iron is converted to rust the remaining iron becomes thinner. This weakens a fence for example and ultimately it may fall over. This happens commonly at the base of a post for example, where there is continual moisture from grass or leaves, and as a result this will be the weakest area.

Chemical changes in food

Take an apple, pineapple, potato or banana. Peel it and cut it into a number of slices. Leave the fruit on a plate or paper towel and observe what happens. After a short period of time a brown colour can be seen on the surface of the fruit. The oxygen in the air reacts with chemicals in the fruit to change the colour. This is an example of a fairly quick chemical reaction. Many people dislike the brown colour and will not eat the fruit as a result. If the fruit was fresh initially, and the colour change occurs because you have recently bitten into it or chopped it, then the fruit is perfectly safe to eat. The change in colour will have no real impact on the taste. However many people dislike the colour change and as a result will throw away perfectly edible food. This is a major problem for supermarkets and they are continually researching ways to keep fruit fresh for as long as possible. In general, keeping the food cold and away from oxygen will reduce any change in colour. Many major fruit suppliers harvest apples for example and store them in very large containers which are chilled and have a very low oxygen level and a high carbon dioxide level. They are then brought to the shelves and we buy them. Many of the apples we buy and eat from the supermarket will be probably six months or up to a year old. This method allows us to eat what we like all year round instead of just during the correct season.

ACTIVITY

Colour change in fruit

You can conduct and experiment with slices of apple for example in an attempt to reduce the colour change. The colour change occurs when oxygen reacts with certain chemicals (enzymes) in the fruit. We can slow this by coating the apple with a little lemon juice.

Slice an apple and separate it into sections. Using a small brush, cover the surface of one of the slices with lemon or lime juice. Place both sections on a paper towel and examine their colour over the course of an hour. You will find one group browning rapidly and the other remaining almost untouched.

Have you ever made a fruit salad? You take a range of fruits, apples, grapes, oranges, berries etc. Peel and chop these as required. Add orange juice or another fruit juice. The juice is slightly acidic and this reduces and slows the oxidation. If the salad is kept in the fridge it can remain clean and edible for days, as you have inhibited the natural chemical reaction that would normally take place. You have also made it taste better!

Have you ever wondered why crisps from a bag are always fresh and 'crisp' yet when you leave them in a bowl overnight they are a little soft or slightly brown? The crisps may have been in the bag for a number of weeks whilst it is stored, transported, put onto shelves etc. yet they are fresh when we open the bag. They stay fresh because the bags are not filled with air but with nitrogen. Air contains oxygen which reacts with the potato. Nitrogen is not very reactive at all. The bags are filled with nitrogen and the crisps stay fresh until you open and eat them. The nitrogen stops the chemical change from occurring.

Why does milk go sour?

This is another example of a chemical change in food. When milk is left in a warm environment for a while, harmless bacteria in the milk react with the lactose in it. The bacteria multiply and create lactic acid. This has a sour taste and smell and leads towards the milk turning to curd. The taste and smell of the milk are good in that they stop people from drinking it. Sour milk is generally not too harmful, but there is a risk that too much bacteria will upset your stomach so it is better to avoid drinking it.

Interesting experiments

It is difficult to undertake some experiments in a primary context as there may be limitations with equipment and apparatus. However there are a number of activities that *can* be done in a primary setting which have the WOW factor and also keep interest and enthusiasm going. Bicarbonate of soda and vinegar are two chemicals which react well together. When vinegar is added to the powder the bicarbonate starts to bubble quite vigorously. The gas given off via the bubbles is carbon dioxide which is harmless. The vinegar can be bought in any supermarket. Bicarbonate of soda has the chemical name sodium bicarbonate (surprisingly) and the active chemical in vinegar is acetic acid.

ACTIVITY

Experiment 1: Making bubbles

You will need a long thin container (like a measuring cylinder), some vinegar and bicarbonate of soda. It's probably better to do this experiment with the container in a small tray in case the bubbles overflow.

Place two teaspoons of bicarbonate of soda in the bottom and then add enough vinegar to about a depth of 2–3cm. The reaction will take place immediately and bubbles will be produced, and this can spread up and over the containers depending upon the size.

We can also use this experiment to show the fire-extinguishing properties of carbon dioxide. Carbon dioxide is denser than air and will stay in the container

(Continued)

(Continued)

once the experiment has stopped bubbling. If we take a lit match and place it at the top of the container the match will go out. This reinforces the point that oxygen is needed for burning. Many fire extinguishers contain carbon dioxide. When used on a fire the carbon dioxide displaces the oxygen and as it is solid or powdery, and is cold, cools the fire as well as extinguishing the flame.

ACTIVITY

Experiment 2: Inflating a balloon – without blowing!

You need a small, empty water bottle, a balloon, vinegar, bicarbonate of soda and a small funnel.

Pour the vinegar into the water bottle up to a depth of about 2cm. Using the funnel, pour about two or three teaspoons of bicarbonate of soda into the balloon. With the head of the balloon pointing downwards slide the opening of the balloon over the neck of the bottle. Try and ensure it is tightly sealed. Tip the head of the balloon upwards and allow the bicarbonate of soda to fall into the vinegar. Leave the bottle to stand and watch as the balloon gradually inflates!

ACTIVITY

Experiment 3: The volcano!

This is a very common experiment and ties in more closely when the class have been studying how the Earth was formed or allied to a study of volcanoes. It is particularly effective if the class has constructed a small model of a volcano out of paper or moulding dough.

If possible leave enough room at the mouth of the volcano that you can place a small plastic bottle in. It is suggested that you possibly construct the volcano around the plastic water bottle. This would give a volcano of a height around 15–20cm. In order to generate a decent lava flow we have to make a small adjustment to the recipe. Firstly, pour some warm water into the bottle (about half to two thirds). If possible add some reddish colouring to the water. This gives the lava a more realistic effect if it is red in a similar way to molten rock itself. Add a squirt of washing up liquid. This helps in keeping the bubbles. Then add about three or four teaspoons of baking soda. Finally, pour in some vinegar and watch what happens. There will be a fairly quick reaction and the lava appears and pours all over the volcano. Again, it is suggested this is done inside a tray, or even better outside in the playground.

ACTIVITY

Experiment 4: Invisible ink

1   *Fruit juice*

There are a number of ways of making an 'invisible ink' which can later be seen by others once they know how to develop the message. A very simple method is using fruit juice as the ink. Lemon, apple and orange juice all work relatively well. Surprisingly, urine also works for this experiment but is not suitable for a classroom environment (although its use in secret messages goes back to roman times).

This is a simple experiment to do as the 'ink' is the juice.

Students need to dip a cotton bud or small paintbrush in the juice and write their message. In general, plain paper or even possibly paper towels would be acceptable. Allow the ink to dry.

Fruit juice is a heat-activated ink in that the secret message reappears after the paper has been held close to a heat source. Holding the paper close to a light bulb or laying it on a radiator will also work. If neither of these is successful, running an iron over the paper will heat it sufficiently for the message to be seen.

This ink changes colour because the acid in the juice is absorbed by the paper and dries. When a heat source is applied it affects the acid more than the paper and it decomposes to leave carbon on the paper. This is what we can see. In effect the heat burns the area where the acid was before it burns the rest of the paper.

2   *Bicarbonate of soda*

The ink in this experiment is a mixture of bicarbonate of soda and water. Add equal parts of water and bicarbonate and stir until it is a smooth liquid. Using a brush or cotton bud they draw or write a secret message and allow the ink to dry.

Obtain some concentrated grape juice. Using a sponge or brush paint over the sheet of paper. The message will then gradually appear. This works because of the chemical reaction between the grape juice and the baking soda. Baking soda is an

(Continued)

(Continued)

alkali and the grape juice is a chemical indicator. It changes colour when reacting with acids or alkalis. This colour change appears where the soda is on the paper to reveal the message.

Summary

This chapter has attempted to describe and illustrate some basic chemical principles that can be investigated in a primary context. The discussion of elements and compounds was designed to give the reader familiarity with the terms which would enable the later chemical reactions to be discussed with more confidence. There is no description of chemical equations as it was felt it was beyond the scope of the text, but the description of some of the experiments – perhaps using words rather than symbols – is perfectly acceptable if a teacher wishes to extend or attempt a more demanding treatment of the topic.

REFLECTION POINTS

- Explain the difference between an element and a compound.

  - An element is a chemical that is composed of only one type of atom. It is pure in the sense that it is composed of only one chemical. It has not been combined with any other chemical. Hydrogen and oxygen are examples of elements as are gold, copper and iron. There are over 100 but many are rare. Compounds are chemicals made up from more than one element. Water has the symbol H_2O (two atoms of hydrogen bonded to one atom of oxygen). Salt is sodium chloride (one atom of sodium bonded to one atom of chlorine: NaCl). Most of what we see and use are compounds as over time the elements tend to react with water or oxygen from the air. Living organisms tend to be quite complex and are composed of large compounds.

- How would we know that a material has reacted and changed?

  - It may have changed colour or appearance. An iron gate will have changed colour and have bubbles on its surface. Burned food will be a charred black colour. Old sandstone buildings may have eroded due to acid rain reacting with the surface.

- What steps can we take to ensure our food remains fresh?

  - We can minimise contact with the air (oxygen); keep food in sealed containers; surround food with a chemical that is not air (e.g. keep potatoes in water).

12.
WATER &
iTS USES

Learning Objectives

Having read this chapter you should be able to:

- understand the nature and composition of the Earth's water
- describe the differences between sea and fresh water and how it can affect us
- outline the processes that we use to treat our drinking water and our waste water
- understand the main processes in the water cycle

Overview

The water in our oceans and streams is the lifeblood of our planet. Liquid water is an absolute requirement in order for life to begin and thrive. Every language has a word for water and no living thing can exist without it. Approximately 71% of the surface of the planet is water, and our seas make up about 97% of the entire planet's water. We drink water, wash with water, cook with water and play in water – though not at the same time! Our ability to build large objects which float in water and can be used to transport us enabled people to explore. This led to the establishment of colonies in far-off places around the planet. It is the management of our water resources that has allowed communities to flourish, and it is its continued management that will sustain our communities in the years to come. In 2010 the General Assembly of the United Nations declared *'the right to safe and clean drinking water and sanitation is a human right that is essential for the full enjoyment of life and all human rights'*.

Water has many unique properties and behaves quite differently from many other liquids. No other material is commonly found as a solid, a liquid, and a gas. It is a light (less dense) gas, a dense liquid, and a less dense solid. When water freezes it increases its volume by around 10%. This makes it less dense than the water, and this is why ice floats on the surface of water.

Composition of the Earth's water

Less than 3% of all the water on the Earth is fresh water. The 'other' 97% or so is (salty) sea water.

Fresh water can also be split into three categories:

Glaciers and ice caps This accounts for about 69% of all fresh water.

Groundwater This is water found underground in cracks and spaces in the soil and rock, and accounts for around 30% of fresh water.

Surface fresh water This is the water we can use for our requirements and it supplies all our water-based needs. It accounts for around 1% of all fresh water. Large amounts of this fresh water are difficult obtain (ground ice, permafrost), and most of the world obtains its fresh water from rivers and streams. Rivers account for less than 0.5% of the available fresh water!!

Historically

Ancient Greek philosophers believed that all matter was composed of four 'elementals': Water; Earth; Fire; Air. It is not correct but it was what was believed to be true at the time. These were described as the classical elements. These elements correspond to more modern concepts of solid (Earth), liquid (water), gas (air) and heat (fire). Much of our science up until Renaissance times was often based on these philosophical ideas. It can be argued that whilst scientifically incorrect, the evolution of modern theories of elements, compounds and mixtures can be traced to ancient models.

Sea water

We all know that sea water is salty. Swimming in the sea inevitably results in our getting some sea water in our mouth and we can then taste it. So how did the salt get there?

The rain that falls from the sky contains some dissolved carbon dioxide. This makes the rain water slightly acidic. When rain lands on rocks in and on the ground it breaks down chemicals in the rocks and carries these along in the water in streams and ultimately to the sea. The most common chemicals released are sodium and chlorine and these are responsible for the salt in sea water. Additionally, when heat from the sun evaporates the water from the surface of the oceans the water vapour is transported away but the salt is left behind. Regions that are hot and do not have a great amount of rainfall are where the sea water is more 'salty'. There are high evaporation rates and little fresh rain so the dissolved salt concentration is high. The Dead Sea is perhaps the most extreme example of this. It is located in a hot central area of the Middle East where the surface water temperature is generally greater than 20°C. Rainfall is very low, averaging around 10cm per year. This leads to a very high salinity level of around 34% and a resultantly high density. People can easily float in the Dead Sea and so it is popular for bathing and the treatment of some skin conditions.

The increased salinity however makes it very difficult for marine life (hence the name) and very few, if any fish live in it. The opposite is true in the North and South

Photo 12.1

iStock.com/RuslanDashinsky

Pole areas. There is not a great deal of evaporation and melting ice is fresh water so salt levels are low.

Sea water for irrigation?

Dry, arid countries can have great difficulty in irrigating crops with limited freshwater resources. An obvious solution would be to use sea water and irrigate areas close to the coastline. Sadly this is not workable at present. The high salt concentration of sea water can affect osmosis (the process by which water enters the plant). This makes it difficult for many plants to take up water and affects their growth and yield. Additionally, over time other salts in sea water can accumulate in the plant roots and become toxic to the plant. There are a number of crops which can cope with the high salt concentration, but more research needs to be done to make this a sustainable way of growing foodstuffs and other crops.

A similar reaction occurs when humans drink sea water. In our bodies our kidneys act as filters. They separate the waste materials from our blood and store them in the form of urine in our bladder which can then be expelled from our bodies. A key factor here is that our kidneys cannot make urine from a concentration of salts of more than 2%. Sea water is generally made up of approximately a 3% salt concentration. If we drink it to quench our thirst our kidneys have to use existing water from our body to reduce the concentration to a level where we can make urine. This dehydrates us and in turn makes us thirstier. Therefore drinking seawater would result in more salt being taken in than being expelled, and a greater amount of salt being left in our bodies.

As sea water is 3% salt and our urine is around 2% we need approximately an extra 50% more water to produce urine. In simple terms, if we drank a litre of sea water we would need at least a litre and a half of water to remove the salt. It is this that leads to dehydration.

What happens to us if we do not drink enough water?

Our bodies use water for a range of purposes and around 75% of our bodies is water. Water is needed for: the production of saliva; the regulation of our body temperature (sweating and respiration); the conversion of food to aid digestion; the delivery of oxygen around our body; the lubrication of our joints; cell growth and reproduction.

The amount of water we use daily will vary depending upon what we are doing, the activities we are undertaking, and our environment. In hot conditions and whilst exercising we can sweat about a litre per hour. If we do not replace the water our body fluids drop and our blood volume and blood pressure may also drop. Dehydration leads to fatigue, darker coloured urine, and if extreme, other potentially fatal complications.

We can survive for a period of a week or two without food, but only a few days without water, and possibly less if temperatures are high.

Fresh water and drinking water

There is a distinction between 'fresh water' and the water we drink from our taps. Drinking water is often referred to as potable which means it is safe for drinking. Fresh water refers to water that is not salty. The water from our rivers will be fresh. It will appear safe to drink and people do take water from rivers etc. when camping or will swallow some water if swimming in lakes, lochs or rivers.

The quality of water in rivers for example is very unpredictable. The water near the surface can be contaminated by bacteria, viruses and parasites. In addition to these the water running off the earth may sweep small pieces of decayed matter, dissolved pesticides and soil along with it. Whilst the water in a stream may appear clear it could contain a number of harmful contaminants.

In order to make fresh water suitable for drinking it has to undergo a number of processes. These processes are:

- collection and storage
- screening
- particle removal
- chemical treatment.

Collection

The rain that falls flows over land into rivers and streams, or percolates through the earth to become part of the 'groundwater'. Many places build dams and vast tunnel networks to collect and store the water in reservoirs. Some reservoirs are natural and others are manmade with large concrete or earth walls built to form dams and hold the water.

Reservoirs serve two main purposes. They act as a store of water to enable us to keep our stocks high even during periods of long dry weather. They are generally kept so that an area or region may have enough water to last for a dry spell of up to 90 days perhaps. They also act as the start of the cleaning process. The water in reservoirs is not fast-flowing. It is relatively calm and this allows small pieces of silt and dust in the water to settle on the bottom of the reservoir. This helps pre-clean the water prior to the treatment process.

Water authorities that supply water for a large area will have a number of reservoirs throughout that area. They are also generally used for sailing and nature walks and such like.

Photo 12.2

iStock.com/kodachrome25

Some areas of Britain however have more rainfall than others. For example, the average rainfall in Glasgow is around 110cm per year with around 200 days of rain per year. The average rainfall in London is around 60cm per year with around 160 days of rain. With a great deal more people living and working in London, the demands on the water supply are much greater and water shortages or reductions in use are more likely to be felt there. England has moved water from one area to another to help during severe droughts, but it is a technical challenge to move millions of gallons of water efficiently.

Screening

This is a simple filtering process but on a large scale. The water in the reservoirs is passed through large metal screens to capture leaves, branches, rubbish etc. that have somehow made their way there. These relatively large objects would almost certainly clog up the treatment process further down the line.

Photo 12.3

'Screen', sustainable sanitation © 2011, used under an Attribution 2.0 Generic license: https://creativecommons.org/licenses/by/2.0/

Particle removal

The process now has to remove the particles that were small enough to pass through the screens.

The first stage is to add a chemical coagulant to the water. This chemical helps small particles in the water to stick together and become larger, which means the sand filters will be more able to trap them.

The second phase is essentially a simple sand-type filter. These are large tanks with layers of fine sand and gravel. The water is pumped into the tank at the surface and is taken away via pipes from the base or bottom of the tank. As the water trickles through the sand, bacteria in the sand act and break down certain compounds in the water and any large particle is trapped by the sand also. The water flowing out of the bottom of the tank is clear and has very little particles held within it.

Disinfection (chlorination)

Chlorine is used as a disinfectant and is added to the water as either a liquid or a gas. It kills harmful microorganisms and can also oxidise unwanted chemicals present in the water. Chlorine is added to the water at the end of the process prior to it being sent to homes via the distribution network.

Some water authorities have added fluoride to the water supply. Fluoride in small doses helps protect against tooth decay, and in areas of high levels of decay where children were receiving large numbers of dental treatments fluoride has been added. It is also in fluoride toothpastes.

In general all the water in our homes is drinkable: it is advised that you use only the cold water from the kitchen for drinking and cooking and that you allow it to run for a little before drinking it (the reason being is that the cold tap comes direct from the mains and has not been settling for any length of time). Hot water can be stored in a tank for some time and may have dissolved tiny amounts of whatever the tank is made of. Lead tanks were used although these are unlikely to be found nowadays, but even in a modern home some lead-based solder may have been used in sealing the joints, so allow the water to run so there is a reduced risk of anything being dissolved in the pipes.

ACTIVITY

Build a water treatment system (sort of)

You will need:

- An empty 2L plastic soft drink bottle.
- Some small pebbles (about 5mm or so in size), coarse sand and some fine sand.
- Alum.
- Filter paper (e.g. from a coffee filter).

Cut the base off the bottle. Turn it upside down and cover the mouth of the bottle with the filter paper. Gently pour in some of the pebbles to a depth of about 7cm. Pour in the rougher sand to a depth of about 2cm. Pour in the fine sand to a depth of about 2cm also.

Fix this filter above a container.

Get some very dirty water, add soil, waste food etc. and mix this together. Shake it until really murky water is visible.

- Add a tablespoon of alum to the water and shake it.
- Let this water settle for 20 minutes.
- Gently pour it into the base of the filter.
- Collect the water that makes its way through the bottle in a container.

This corresponds to the collection, storage, screening and particulate removal processes outlined earlier. Alum acts as a coagulant and binds larger particles together and so they sink to the bottom or are trapped by the sand and pebbles. You should now have a container with water that is very clear and has little or no particles in it. It is however *not* fit to drink – it has *not* been disinfected, but we have simulated most of the processes our drinking water undergoes.

To see video clips of how to build a water treatment system go to the companion website for the book at
https://study.sagepub.com/chambersandsouter

Our own waste water

When we flush the toilet, empty the sink or take a shower, the water we have used is not safe to pour into our drains. It contains our own waste products such as faeces and urine, plus it may also contain grease, dirt, detergent, food waste and so on. This water is heavily contaminated and needs to be treated before we allow it to return to the seas and rivers. Our waste water is taken from our homes via an arrangement of sewer pipes. This system is generally separate from the drains along the side of the road. Drains are intended to allow rainwater to flow away in order to keep our roads and pavements clear. You should never pour waste water into a drain.

The first stage is similar to our drinking water system and is a large screen or mesh which catches anything which could block the sewers or damage them. Screening generally stops nappies, face wipes, sanitary items etc.

The second stage is to allow the sewage to settle. It still contains organic solid matter and the waste gradually sinks to the bottom of large settlement tanks. This waste is called sludge and gathers in the bottom of the tank. These tanks are often circular and large arms scrape the surface and guide the sludge towards the centre of the tank where it is pumped away.

The next stage is to remove the smaller bugs and chemicals. Some tanks have air pumped into them and this aeration encourages good bacteria to break down the harmful stuff.

The final stage is another settling process and the clean water at the surface is returned to the water system.

The sludge

The nasty, solid sludge is generally recycled and farmers can use it as fertiliser. It can also be used in specially adapted power stations where the sludge is burned or used to create biogas which is burned to generate electricity.

There are a number of other processes which can be used in the treatment of sewage, but the processes outlined above are the main features of an average sewage treatment plant.

Drinking water across the world

The water that we drink in industrialised countries is a luxury that is perhaps not fully appreciated. Over 6,000,000,000 people have access to relatively fresh drinking water. Not all of it will have come via the processes outlined above. Some will be from wells which draw water from the groundwater supply. This is moderately clean water as the rocks underground can act as filters and the water is not affected by human waste or bacteria to a large degree. It is however not as clean as the water drunk in Britain and there are some disturbing statistics. The World Health Organisation (WHO) states:

- 663 million people rely on basic, traditional drinking sources, with 159 million taking their water from lakes and rivers.
- 1.8 billion people use drinking water which has some form of faecal contamination.
- Contaminated water can transmit diseases such as diarrhoea, cholera, dysentery, typhoid and polio. This water is estimated to cause over 500,000 diarrhoeal deaths per year.
- By 2025, half the world's population will be living in water-stressed areas.
- Some 240 million people are affected by schistosomiasis – an acute and chronic disease caused by parasitic worms found in contaminated water.

The countries with the lowest access to fresh or improved water are mainly impoverished countries which have undergone major unrest. For example, Somalia, Ethiopia and Afghanistan all have very limited access to fresh drinking water. This lack of fresh water leads to the spread of disease and as a result the life expectancy of Somali adults for example is just over fifty years. In Britain the industrial revolution led to large groups of people living and working in cramped, unhealthy environments. Water and sanitation were not able to meet the demands of the population and many people were dying. In the 1880s our water supply and sanitation network needed to be upgraded. Fresh water was brought into cities from relatively clean stores outside the industrial areas. A comprehensive sewage system also took away the waste from heavily populated areas. This led to an increase in life expectancy of fifteen years in a space of approximately four decades.

The water cycle

The water cycle or the hydrological cycle is the name we give to the description of the various processes which cause or contribute to the ways in which water is spread, distributed and moved on the planet. In general the amount of water on Earth remains constant over time (it cannot fly off into space for example), but the various sections such as sea water, fresh water, atmospheric water etc. may fluctuate slightly.

Water in the sea can evaporate and form clouds or make the atmosphere very humid; it can fall to Earth and land in seas, rivers or on land. It may be frozen as ice or leak underground into groundwater. Evaporation, transpiration and condensation all involve energy being transferred and this transfer of energy leads to temperature changes. For example, when water condenses energy is 'released' into the atmosphere, causing an increase in temperature, but when evaporating it takes in 'energy' slightly lowering the temperature. You will feel colder when wet as the water is taking energy from your skin in order to evaporate. That is why we sweat. This gradual exchange of energy affects the climate as a result of the weather. The process is hugely complex and the computers required to operate weather-modelling programmes are hugely powerful.

The Sun is the cause of the water cycle. It heats water in the seas and this evaporates as water into the atmosphere. Water vapour can also enter the atmosphere from leaves via a process known as evapo-transpiration. Water vapour is less dense than nitrogen and oxygen and will rise through the atmosphere. As the vapour travels to greater heights the temperature and pressure drop and this causes very small droplets of water to form. These droplets condense to form clouds and fall very slowly downwards due to their size and air friction. They can, however, be carried upwards by rising air and this is what makes clouds appear to float. Clouds are essentially large amounts of tiny droplets being kept in the air by updrafts and thermal currents. If these droplets condense at low levels we have fog.

Over time and with the correct atmospheric conditions, these droplets accumulate and become larger and then fall as precipitation. Rain, hail, sleet and snow are all caused in this way. Cloud seeding is a way to modify the weather patterns and to cause precipitation. Aeroplanes – or in some cases rockets – release a fine powder (dry ice for example) into already existing clouds. The ice particles act as 'seeds' for the vapour to condense on. This vapour then condenses and the particles become larger, ultimately falling as rain or snow. Many large ski resorts use this method to generate new snow for the runs.

You can demonstrate this idea of cloud formation in the classroom. It is tricky but easy to repeat once you get the temperature correct.

ACTIVITY

Making a cloud

You will need a reasonably large glass jar, a small plate as a lid, and some ice.

Step 1

Fill the jar about half-way full with very hot water. (Water that has been boiled and then left for about a minute is about right.) Cover the jar with the plate or lid and place ice on top.

(Continued)

(Continued)

Step 2

Let this sit for a few minutes – you will start to see a cloud form. You can observe convection currents as the hot air rises to meet the cold air and then sinks again. This can be used to reinforce the idea that the water in our seas evaporates and rises to form clouds.

Lift off the plate and watch the cloud gradually spread out and disappear.

You can repeat the experiment using an aerosol spray which can be used to replicate the cloud-seeding idea. Repeat Step 1 and then lift the lid and spray a small amount of 'spray' into the jar. The aerosol acts as a 'seed' and the moisture condenses around the droplets to form a thicker cloud. It won't rain sadly, but it's not a bad experiment.

The UN figures state that across the planet 814mm of rain falls on land surfaces each year. About 60% of this is returned to the atmosphere by evapo-transpiration and the remaining 40% is available for human use. This equates to about 16,000 litres per person per day!

However, this water is not distributed evenly in geographical terms, and much of it is not accessible. It is interesting that people in Iceland average 1,400,000 litres per person per day, but people in Kuwait receive 16.

This process occurs all over the planet. At some point on the surface there will be rain, snow, evaporation, cloud formation and so on. The process never stops and this transfer of water and energy is constant.

Solid water

Most people will be familiar with liquid water and how it behaves and is used. The Celsius temperature scale is a universal scale based upon the temperatures at which water freezes and boils. We define 0°C as the temperature at which water freezes or ice melts and 100°C as the temperature at which water boils. Solid water as ice or snow is interesting. Children will be familiar with ice as being cold and slippery and snow as powdery and suitable for building snowmen. In cold weather water turns to ice. Water vapour turns to snow in cold environments, but the formation of snow relies upon very small particles of water vapour to freeze and then come together. Snow could be thought of as tiny specks of ice aggregated together. This is what gives snow its white appearance rather than the transparency of ice. When white light from the Sun enters the snow it travels through millions of little crystals and is reflected every which way there is: it then reappears and reaches our eyes as white. Ice is clearer because of the crystal structure of the atoms, but it too can appear opaque if there are cracks in it.

An interesting experiment is to investigate how much water expands when it changes into ice. This can be done with small jars – drawing a line around the jar at some point, filling it with water, and then placing it in a freezer. Additionally you could fill a jar or an ice tray to the very brim and then place that in the freezer. You will find that the ice will sometimes move up and over the lip of the jar or tray but will not spill. This is because the ice freezes from the top of the container and as it expands the section that does so is already solid. This leads to some strange phenomena of ice spikes appearing in ice trays or occasionally on a bowl left outside when cold. These occur because the ice freezes from the outside. As this expands it creates pressure in the centre and the only place for the liquid to go is upwards where it freezes.

Photo 12.4

iStock.com/DonNichols

Icebergs

Icebergs are massive lumps of ice that break off the ice shelf and float away in the oceans. As they are mainly water they will be less dense and float. Approximately 8–10% of an iceberg will be above water and as such they can be hazardous for shipping.

ACTIVITY

Iceberg investigation

Get a reasonably large bowl and a large lump of ice. Place the ice in the bowl and fill the bowl to the brim with water. Ensure the ice is floating. You should then have a bowl with a large lump of ice, with a fair bit of it protruding above the water level.

(Continued)

(Continued)

Ask learners to predict what will happen to the water level in the bowl when the ice melts.

The answer should be that the water level will remain constant. As the ice melts it reduces in volume and will not spill over the bowl. This essentially means that a melting iceberg will not increase the water level. As it is already in the water it has raised the water level. Therefore when it melts it will not increase the level.

ACTIVITY

Making frost

Get a metal container and fill it with ice and three or four tablespoons of salt. As the salt combines with the water it lowers the temperature to a few degrees below freezing. Water vapour will condense on the side of the can and then freeze. If left for a while (out of a warm area) it will develop a skin of white frost around it.

Summary

This chapter has introduced some of the ways in which water is used in our planet. Water is a unique chemical as regards many of its properties, and it is this uniqueness that makes it so essential. Clean, fresh water is crucial for our existence, and much of the planet does not have access to it. How we use and treat our water has led to our general well-being and as a result it is possibly being taken for granted. Our water companies however do not take it for granted and they are continually seeking ways to ensure our supplies are fit for purpose. It should be emphasised that our water is a resource and that it is how we manage it that ensures our current, healthy position.

REFLECTION POINTS

- Describe how a cloud forms.

  ○ Heat from the Sun will cause moisture on the ground, sea and elsewhere to evaporate. This gaseous water will rise and when the vapour reaches an area of the atmosphere, where the temperature is less for example, small droplets will form. These droplets may then come together and become larger, ultimately leading to darker clouds and possible rainfall.

- Why is it unhealthy for us to drink salt water?

  ○ In our bodies our kidneys act as filters to separate and remove waste products. They store this waste liquid as urine in our bladder. The process that does this is dependent upon the salt concentration of the waste. Sea water has a higher salt concentration than our kidneys can deal with. We have to reduce this concentration before it can be removed. This process removes water from our body in order to do so. This makes us more dehydrated and unwell.

- What actions could we take to conserve our water supplies?

  ○ We can reduce our usage. Many toilets now have two settings (4 litres and 6 litres) when we flush. Use the small setting wherever possible. Do not use the toilet for getting rid of paper for example. Do not spend 15 minutes in the shower. Get in, get washed, rinse off, and get out. Full dishwashers are better than washing by hand and save water. When using drinking water some people let the tap run every time they want a drink. Let it run once, then fill a jug and place it in the fridge. Check for leaks. In England and Wales more than 3 billion litres are lost (in the water industry) per day due to leaks in the pipe and reservoir infrastructure.

Learning Objectives

Having read this chapter you should be able to:

- understand our place in the solar system and how the Sun is essential for our existence
- be aware of early explanations of planetary science
- understand how calendars developed and what is meant by a day and a year
- explain the seasons and tides in relation to the Sun and Moon

Overview

The universe is the name we give to all of the matter, space and energy that exists and has existed. It contains all the stars, planets, galaxies, constellations, space dust and other objects that we can see and observe with our telescopes. Conceptually it is difficult to appreciate that there is not an 'end' to the universe or that it is not contained within some sort of frame.

We can observe planets in our local vicinity on a clear night using a simple set of binoculars or the naked eye. The Sun is easily visible and supplies our planet with the energy it requires to sustain life. It is the basis of our solar system and its gravitational attraction and radiated energy dominate the area of space around it. The motion of our planet through space and around the Sun is regular and leads to changes in the length of our days.

Looking further we can observe stars which form certain shapes when viewed from Earth and these shapes were given names and used to assist in navigation. Farmers

Photo 13.1

iStock.com/danmitchell

could determine when to plant crops by measuring the length of the day. The shortest and longest days of the year give a good indication of the seasons as do the spring and autumn equinoxes.

The image in Photo 13.1 is of Orion which can be seen clearly with the naked eye. It is visible throughout the world and is more prominent at certain times of year.

The bright dots are stars and to us it appears these are in relatively close proximity to each other and form a certain shape, but this is not the case. The stars are separated by great distances and some are many times further away than others. To our 'view' however they appear to come from the one, narrow area.

Beyond these constellations (and only really visible with large telescopes) are enormous, massive groups of stars. These aggregate together due to gravitational attraction and form galaxies. These galaxies form certain shapes and the images from telescopes show some remarkable formations.

Photo 13.2

Credit: European Space Agency & NASA

Photo 13.3

iStock.com/ManuelHuss

The images in the photos show examples the shapes galaxies can form.

Our own star is one of the bright 'dots' in a galaxy similar to the spiral shape shown. Each galaxy contains around 100,000,000,000 stars, and each star could have a number of planets orbiting around it. Our galaxy is called the Milky Way and can be seen with the naked eye on a clear night.

Photo 13.4

iStock.com/ MarioGuti

It is difficult to determine but our universe may contain over 200,000,000,000 galaxies. The sheer scale, size and numbers involved in describing the universe make it difficult for teachers to give indications of scale beyond that of the solar system, but it is part of the universe we inhabit.

Our Sun

The Sun is our nearest star and is our supplier of energy. All the energy the Earth needs to maintain a supply of liquid water, keep our environment at a reasonable temperature, ensure plants grow etc. comes from the Sun.

The Sun was formed over 4.5 billion years ago from a very large cloud of molecules of mainly hydrogen and helium. It is thought a shockwave from a nearby exploding star would have compressed some of the molecules closer together, and that they would have started to gather or collapse in a smaller region because of the gravitational attraction of the molecules (for more on gravitational forces see Chapter 19).

As the cloud started to collapse and 'fall in' on itself, it formed a rotating disc with most of the matter concentrated in the centre. The actual physics of why it formed this shape is beyond the scope of this book, but it is a combination of moving matter and being attracted to a central region and as such it is extremely complex.

Eventually the matter at the centre became so concentrated, due to gravitational effects, that the pressure increased at the core, eventually triggering the nuclear reaction that is the Sun.

Photo 13.5

iStock.com/hadzi3

The matter at the more distant edges of the cloud gathered together to form planets and other objects such as asteroids.

The energy emitted by the Sun is a result of nuclear fusion. Smaller atoms (hydrogen) combine to form larger atoms (helium) and as a result there is a small amount of matter that is surplus. This matter is converted to energy and emitted by the Sun. It converts about 700 million tons of hydrogen each second, and in its 'life' so far has used about half of its supply of hydrogen.

The helium created then undergoes other transformations and can combine to form larger and heavier elements. The carbon and other elements in our bodies were created in some stellar reaction. (We are all made from stars!)

The majority of the energy emitted by the Sun is in the visible band. It also emits significant amounts of energy in the ultraviolet and infrared bands, and smaller amounts in the radio, gamma-ray, x-ray and microwave bands. The ultra-violet radiation from the Sun is what leads to people's skin going darker, sunburn, and possibly skin cancer.

The solar system

A simple definition of the solar system is that area of space around the Sun where it is the dominant influence in terms of energy and gravitational attraction.

For teaching purposes with primary schools it would be appropriate to concentrate on our Moon, local planets, and more familiar space objects. It is a common misconception that our solar system stops at Uranus or Pluto. The solar system extends far beyond the realms of the planets and dwarf planets such as Pluto that we may be familiar with. There are objects which orbit the Sun but are so far away that our local solar system is tiny in comparison.

The local solar system

The four inner planets (Mercury, Venus, Earth and Mars) are hard, rocky planets which are in relatively close proximity to the Sun. The distance from the Sun to Mars for example is smaller than the distance between Saturn and Jupiter.

The four outer planets (Jupiter, Saturn, Uranus and Neptune) are larger gaseous planets. The size of these 'gas giants' makes the four inner planets seem very small. Jupiter is over 300 times the size of the Earth, and is so large that its famous red spot is greater than the size of the Earth.

Photo 13.6

iStock.com/ChrisGorgio

The photo shows a representation of the planets and the Sun and gives some indication of these as they orbit the Sun. It is incorrect however. The scales depicting the size of the planets and the distances between them are wrong. Indeed they are incorrect in almost every image you can find. The reason for this is that the distances are so great relative to the size of the planets that a true representation is difficult.

ACTIVITY

A true representation of the scale of the solar system

This can really only be done with the four inner planets (Mercury, Venus, Earth and Mars).

Ask the class to make three small balls of plasticene or play-doh. (It is better if you roll these with blue plasticene and then add thin layers of green plasticene to represent the land masses.)

Make these 1cm in diameter. The inner planets are of a roughly similar size so minor variations in diameter will not matter.

At this scale the diameter of the Sun is around 110cm. If possible get a circular piece of card or possibly a circular table about this size and place it on its side at one end of the room. (Sadly Earth is now about 117m away. This is longer than the school's football pitch.)

Make another ball of about 4mm diameter and place it 30cm from the Earth. This is our Moon. (At this scale Mars is a further 60m or so away.)

An assumption made here is that the planets are all aligned (i.e. they are all in the same general direction). This is not normally the case. Mars could be 60m further out from the Sun, but it could be some 200m away on the other side of the Sun!

This true scale representation is extremely difficult for many people to appreciate, but it does highlight the sheer scale of the solar system and the virtual impossibility of manned travel to other planets.

Our planet receives radiated energy from the Sun; it absorbs some and reflects the rest. It reflects about 40% of the energy it receives. This energy from the Sun allows the atmospheric conditions we are familiar with to be maintained.

We have developed because our planet has a stable environment, plenty of liquid water, and a benign atmosphere. If the Earth was a small distance closer or further away from the Sun, water would not exist in liquid form and very probably neither would life. The region of space the Earth is within is referred to as the 'goldilocks zone' because it is neither too hot, or too cold, and therefore capable of sustaining life.

Satellites

Satellites are generally defined as objects which orbit planets due to the gravitational attraction of the planet.

There are many natural satellites which orbit the planets, and we call these moons. Our own Moon is believed to have been caused by a large meteor colliding with the Earth and the dust from that collision coming together to form the Moon. There are also many icy or rocky objects that can be trapped or locked around us by gravitational attraction. One of the features of large planet-like objects is that they are held together by their own gravitational attraction. This attraction is very great with large objects and as they were formed from matter coming together the attractive force pulls the matter together equally in all directions. This 'distributes' the attractive force equally in all directions, meaning that most large planetary objects are almost spherical. Smaller asteroids are not large enough for this to happen and have irregular shapes. Indeed one of the criteria for being considered a planet is that it must be reasonably spherical.

Early astronomy

Astronomy is arguably the oldest of the natural sciences. Various cultures have for many years given huge significance to celestial events, and the prediction of when events such as the winter solstice or eclipses occur was given great spiritual importance. This is essentially the basis of astrology – the linking of events related to the position of the stars and planets in order to explain their influence or impact on events on Earth.

Remember here that astrology is not a science, it is a belief system. People choose to believe that the position of the stars when they are born will somehow affect them. There is no link whatsoever with the alignment of the stars and your character or personality.

Early cultures were very superstitious and believed in gods and deities who were linked to the stars. They relied heavily on celestial events for various ceremonies and any astrologer who predicted incorrectly the time of an event would often be punished or executed.

This required detailed records of the passage of stars, the Moon, and such like. There are cave drawings and markings indicating lunar cycles from over 20,000 years ago. Astrologers (as they were at that time) observed the movements of stars and other objects across the night sky. It was recorded that some 'stars' moved more quickly than the majority of the other stars. These were given the name wandering stars, the Ancient Greek word *planetes*. Early astrologers could identify five planets, the Moon and the Sun.

Table 13.1

Astronomical body	Roman god	Greek god	Hindu god
The Sun	Sol/Apollo	Helios/Apollo	Surya
The Moon	Luna/Diana	Selene/Artemis	Chandra
Mercury	Mercury	Hermes	Buddha
Venus	Venus	Aphrodite	Shukra
Mars	Mars	Ares	Mangala
Jupiter	Jupiter	Zeus	Guru/Brihaspati
Saturn	Saturn	Cronos	Shani

The accurate observations required for predicting events meant people recorded the slow movements of stars and planets across all sectors of the sky. They also recorded their motion over extended periods of time to provide astrologers with the information they needed to forecast eclipses and the like.

The planets (in astrological terms) were linked to gods across different cultures:

Shakespeare, among others, used the public perception of stellar events in his plays. Calpurnia, Julius Caesar's wife, was concerned that comets had been seen and begs her husband not to go out on the ides of March:

> When beggars die there are no comets seen; The heavens themselves blaze forth the death of princes.

Audiences at that time would have also believed in the importance of celestial events, and this would have linked to their fears and superstitions to give more emphasis.

The earliest recorded information regarding the observations of stars and planets and their positions and movements comes from Babylon in modern-day Iraq. There are also records of observations from Chinese, Central American, and North European cultures.

From about 600 BC ancient Greeks were perhaps the most astronomically advanced. They used what data they could take from the Babylonians, and possibly others, and attempted to explain the motions of the stars. Eclipses could now be predicted.

Eratosthenes (276 BC–194 BC) was a Greek mathematician and astronomer. By observing the shadow cast by the Earth on both lunar and solar eclipses it was known that the Earth was circular, in the shape of a sphere. By measuring the length of shadows at two different points, at the same time (Alexandria and Aswan in Egypt), he calculated a value for the circumference of the Earth which was remarkably accurate.

Early planetary theories

The geocentric model

This model, first developed by a Greek astronomer Heraclides, places the Earth at the centre of all things with everything else rotating around it.

This model was proposed around 400 BC and was accepted generally until approximately the seventeenth century. Ancient Greek scientists drew up a set of rules – some quite complex – which described the motion of the planets and allowed us to make calculations with some accuracy. These were formalised by Ptolemy in the second century BC and they lasted well. All the bodies in this system moved in perfect circles around the Earth. It had to be around the Earth as we were the centre of everything. There were some issues with this model but it explained most things and was accepted.

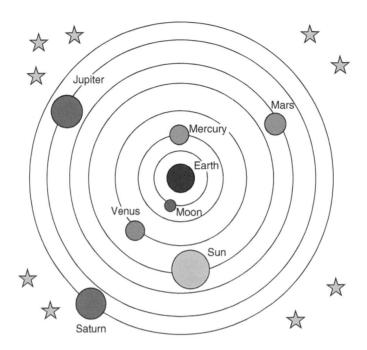

Figure 13.1

The heliocentric model – the Sun at the centre

This model, first put forward by another Greek astronomer Aristarchus, placed the Sun at the centre of the solar system.

This was proposed at about the same time as the geocentric model but there were some difficulties with this being accepted. It had the Earth orbiting the Sun but the people on Earth had no perception of this motion. Their life was pretty stable in movement terms.

The Earth's orbit

How does the orbit of the Earth around the Sun affect our life? The answer is, quite dramatically.

We use the movement of the Earth, the Moon and the Sun to measure the passage of time. These periodic motions define terms such as a day, a month, or a year. However these definitions are not quite as simple as they first appear:

- The time for the Earth to make one rotation on its axis is one day.
- The time for the Moon to make a complete orbit around the Earth is one lunar month.
- The time for the Earth to make a complete revolution around the Sun is one year.

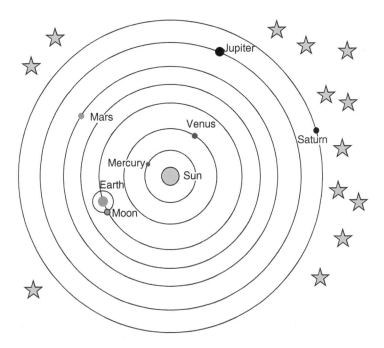

Figure 13.2

More importantly, they are not measurable in simple multiples of each other. By this we mean that a whole number of days does not make a lunar month, and that a year is not measured by a whole number of days. The period of these orbits and rotations does not fit into nice, simple, easily numbered slots.

This means that we have to make adjustments as time passes, such as a leap year for example. This then helps keep our calendars regular.

What is a year?

In simple terms it is the time taken for a planet to orbit a star. Our year is the time taken for the Earth to complete one full orbit of the Sun (i.e. this is our 'Earth year'). There are slight differences in what certain calendars regard as a year. We regard January 1st as the start of the New Year but this is just an arbitrary day. Would it not seem more reasonable to commence the beginning of the year on the longest or short-est day, or one of the equinoxes where we have equal hours of sunlight and darkness? The modern Iranian calendar for example starts and ends on the vernal equinox on or around March 21st each year.

For school-teaching purposes we will define a year as the time it takes for us to return to the same point in space relative to the Sun. This doesn't actually turn out to be a whole number of days (365.24 days) so we have to make a small adjustment every so often. We do this by adding a day every four years. This then allows us to return to pretty closely the same point, but even so we will still have to make other adjustments.

Our orbit and the seasons

We have different seasons throughout the year because the Earth is tilted at approximately 23°. This means that at certain times in the Earth's orbit round the Sun the Northern hemisphere will tilt towards the Sun and the Southern hemisphere will tilt away from it.

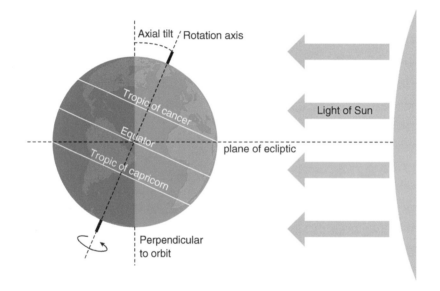

Figure 13.3

Source: The Met Office (http://www.metoffice.gov.uk/learning/learn-about-the-weather/how-weather-works/seasons)

In Figure 13.3 the Northern hemisphere is tilted towards the Sun. As the Earth rotates on its axis the Northern section gets more hours of sunlight shining on it each day. This means the temperature gets warmer and plants and crops can grow. It is advisable to plant our crops or have our farm animals born just before the weather starts to improve so that the increase in energy and temperature allows them to thrive.

The section at the very top is always in sunlight. This occurs in places such as northern Finland and Iceland. It is referred to as the 'midnight sun', and areas such as these can have continuous days of sunlight depending upon how far north you are. Conversely, the southern section is receiving less hours of sunlight each day and so it is winter there.

When the Earth is at the other side of its orbit around the Sun, the hours of sunlight are reversed and it would be winter in the north and summer in the south. Additionally, the area that was in total sunlight will now also be in total darkness! This is sometimes referred to as the 'polar night'.

What is a month?

A month is generally the period of time it takes for the Moon to orbit the Earth. The word 'month' is derived from 'moon'. It takes approximately 29.5 days for the Moon to orbit the Earth and there are just over 12 lunar months in a year. This makes calendars difficult to operate if they are based on the lunar cycle. Each 12 months is 5 days short of a correct year, so our dates gradually move out of sync with the seasons.

The Moon orbits the Earth due to our gravitational attraction, but it also attracts objects on the Earth. Matter 'facing' the Moon will experience an attractive force, and the water on Earth in that position will rise towards the Moon. This leads to high tides in that area and to low tides in the water which is perpendicular to the Moon. The water furthest away from the Moon also rises, but this is due to a combination of the Earth's and Moon's rotation. This is too complex to explain here, but the result of this is that we have two 'high' tides per day, on either side of the planet, caused by the Earth's and Moon's rotation.

Length of days

Look at point X on the right-hand side of Figure 13.4. Sunlight shines on the lighter shaded area. As the Earth rotates, the point X receives sunlight for a shorter time. The darker shaded area is equivalent to night-time. The time the Earth is in this region is longer than for the lighter shaded region. This equates to shorter days and longer nights.

A common misconception here is that the Earth is closer to the Sun in summer and further away in winter. This is untrue. If that were the case it would be summer for the whole planet when closer to the Sun, and winter for the whole planet when further from the Sun.

As the Earth orbits the Sun the length of day varies. Our shortest day occurs when our tilt is aligned away from the Sun (i.e. the winter solstice). This shortest day occurs at the same position and time every year and is thus a good way to record a year. When we are at the other side of the Sun our tilt is towards it. This is our longest day of the year (i.e. the summer solstice). When the Earth is halfway between the summer and winter solstice, the length of day and night are the same. This is the spring and autumn equinox. These four days are very good points to measure the time of year and the onset of the seasons. They occur at the same point of the orbit each year.

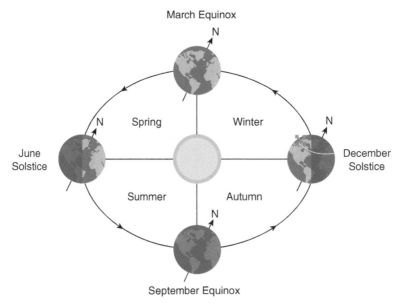

March Equinox

Spring

Winter

June
Solstice

December
Solstice

Summer

Autumn

September Equinox

Figure 13.4

Bakhtiar Zein © 123RF.com

So why can't we see the stars during the day? Light from the Sun reaches our atmosphere and is directed downwards to the surface. Blue light is directed more than any other colour and this is why our skies are blue. The light from the stars still reaches us, but it is so much dimmer than the sky that we cannot see them.

ACTIVITY – LENGTH OF DAY

Temperature and length of day

Measuring the hours of daylight is a relatively simple activity but difficult when daylight begins and ends outside of school time. A possible solution to this would be to take the length of daylight from some newspaper or website. Note the length of the day and the temperature at noon for a particular day – Monday for example. Note the number of hours of daylight and plot this on a graph, and then let the class observe the gradual increase and decrease as the seasons change.

This will give a nice distribution curve. Additionally, the class could add some of their own data. The temperature outside the classroom could be measured and added to the graph in some way. This would, over time, give an indication of the relationship between hours of daylight and the temperature outside the classroom. There will obviously be variations within this, but a qualitative relationship between the two could be made, and comparisons with charts from holiday brochures or almost any weather website could also be made using their own data. The Met Office provides such graphs for most places.

Interesting points could be raised here. The temperature in August for example may continue to rise, even though the number of daylight hours decreases. There could be discussion around the atmosphere still getting more sunlight, and as the duration of that sunlight is falling why there is still enough to cause the air temperature to rise. This will also allow for a general discussion with regard to cold/hot days/direction of wind etc.

ACTIVITY

The Sun over the course of a day

This is best done by asking a pupil or pupils to sit at a set point in the classroom (a 'Sun Seat'?). At hourly intervals, look at the Sun through the classroom window and note the position of the Sun *on* the window, perhaps placing a large piece of card or tracing paper on the window whilst this is being done. Depending upon the time of year you will create a 'curve' something like that in Figure 13.5.

Figure 13.5

This could be drawn on the 10th of December, March and June to give real data regarding the position of the Sun.

ACTIVITY

The shadow stick

This activity allows children to plot the path of the Sun using a stick and the shadow it casts. Select a point away from a building which may cast a shadow.

(Continued)

(Continued)

Insert a stick (~ 1m long) into the ground and have it pointing vertically upwards. Mark the position of the top of the shadow on the ground. This will 'draw' a shape similar to the previous activity but inverted. A small shadow will appear when the Sun is highest.

ACTIVITY

The Sun over the course of a year

This is more complex and less dramatic in that the results take a long time to be seen. This activity involves plotting the position of the Sun at a set time (e.g. midday). The key is to ensure that the position is recorded at the same time.

Record the position of the Sun (at midday say) and repeat this using the 'Sun Seat' every couple of weeks. Try and do this for a full year although the results will be noticeable even after a few months. The shape the Sun traces over the course of the year is a 'figure of eight'. It is called 'an annulus'.

Photo 13.7

J Fishburn, 1998 © Reproduced under the terms of the GNU Free Documentation License.

The shape will vary depending upon our position on Earth but in general will be similar to that shown.

Summary

This chapter attempted to explain the position and motion of our planet within the context of the solar system and space. Our motion – relative to the Sun and Moon – leads to the explanations of everyday physical occurrences such as the length of the day, seasons and tides. These are complex and not easily explained to young people. The sheer scale and speed are difficult to comprehend. We rotate on our axis once per day. This equates to a speed at the equator of around 1000mph. In addition our speed around the Sun is roughly 67,000mph. We are also rotating around the centre of the galaxy, yet we feel almost nothing. However our fascination with 'what is up there?' provides a large and positive motivational tool for this topic, and our realisation of the enormity of the solar system often sustains interest in our making observations of the night sky. Additionally, it is this observation of the stars, Moon and Sun across the sky that supports our understanding of our place in the universe.

REFLECTION POINTS

- How could you explain or justify to a class that the length of a day is related to the rotation of the Earth?

  - One method would be to use a model globe of the Earth. Set this in the middle of the classroom with a light bulb across the room used to simulate the Sun. Have the globe with Great Britain facing the Sun (this is daytime). Rotate the globe until Britain is away from the Sun (this would be nighttime). For older children rotate the globe slowly until the light shines just on Britain. Continue rotating and to an observer on Britain the Sun will appear to move across the sky from left to right. This could then be matched with a view of the Sun during the day. Learners would see the Sun move across the sky in a similar way to your explanation.

- Why can you sometimes see the Moon and the Sun in the sky at the same time?

  - The Moon orbits the Earth every month. There are occasions when the light from the Sun reflects off the Moon and reaches us. If the Sun is across the sky from the Moon we can see both.

Figure 13.6

(Continued)

(Continued)

- Why is a calendar based on lunar cycles not accurate?

  ○ A lunar calendar based on 12 lunar months only lasts for 360 days. Every year, the first day of a new year will be five days out. If one year started on the 1st of October say, the same day the following year will be the 25th of September, then the 20th and so on. After six years the calendar would be out by a full month. Many religious calendars are lunar based but in countries that use these for religious festivals they also operate a different 'official' calendar.

14.
ENERGY

Learning Objectives

Having read this chapter you should be able to:

- understand what the concept of energy means
- be aware of the difficulties in explaining and describing events using energy
- have a good knowledge of how energy in the form of heat is transferred
- be aware of misconceptions relating to explanations involving energy
- explain simple electrical circuits in terms of energy moving

Overview

The term energy is used widely in a great deal of circumstances. It is possibly one of the most overused words today and not surprisingly means different things to different people. Some common uses of the word and associated terms are:

- energy suppliers (who don't actually sell energy!)
- energy foods and drinks
- renewable energy
- energy sources
- energy crises (occasionally)
- Department of Energy and Climate Change (DECC).

As teachers of young children, a major difficulty for us arises when we are attempting to explain and describe a concept such as energy in a meaningful way whilst not reinforcing incorrect science or using some of the very bad or wrong science in this area. Even the unit we measure energy in is generally incorrect. Many people refer to calories when referring to the 'energy content' of certain foods. The scientific unit for energy is the *Joule* (J). This unit was named after an English physicist, James Prescott Joule, who came from the Salford area. He was the son of a wealthy brewer but was fascinated by machines and electricity, and spent much of his life investigating what machines could do and how they operated.

His work involved attempting to quantify the amount of 'work' that machines did. Essentially he was trying to develop more efficient machines in an attempt to extract the maximum amount of 'work' possible from burning a pound of coal for example. This was possibly the beginning of the concept that energy could be converted from one form to another. Joule tried to obtain a value for the 'mechanical equivalent of heat' (i.e. how much energy – the mechanical turning of a machine for example – was required in order to heat a certain volume of water). He obtained a very specific value for the amount of 'work' in these experiments and all were found to be roughly the same. This was leading to the idea that the same amount of work resulted in the same temperature rise. One experiment involved a container of water with paddles. These paddles were attached using a rope and connected to a weight. As the weight fell it pulled the rope which then spun the paddles inside the water. This mixing or stirring increased the temperature of the water and Joule equated the 'work' done allowing the mass to fall with the increase in the temperature of the water. The paddles stirring and agitating the water were thought to be transferring something (energy) to the water, thereby increasing its temperature.

The unit *calorie* was defined as the amount of energy required to increase the temperature of 1 gram of water by 1 degree Celsius. This is a very small unit and the more accepted unit that we can read about in food labelling refers to the energy required to increase the temperature of 1kg of water by one degree. This is generally the definition of the calories you will see printed when looking at a food wrapper or label.

This common, everyday use of the word 'calories' in newspapers and on television causes difficulties for teachers. The unit has no meaning outside the food and dieting industry. The public programmes on this topic also often contain some very bad and incorrect science. Many diet advertisements will often contain claims from some study but these are difficult to pin down. Even the language associated with energy and foods leads to misconceptions with children and the public.

'Oh, look at the number of calories in that cake!'

'I need to exercise to burn off those calories!'

The phrases themselves are not bad or an attempt to mislead young children, but they do imply those 'calories' can move from the cake and then be burned off by doing some activity. The reality is that it's just a tad more complex than that.

Children's views of energy

Children's perceptions of energy and its uses have proven an interesting topic and much has been written about this. There is a wealth of literature in this area that can be accessed using a couple of simple search terms. A number of researchers, in trying to categorise or impose some form of order in this area, have advocated categorising the general framework of how children use the term (see Sefton, 2004) in everyday explanations. In broad terms the classification contains concepts such as the following:

- Energy is associated with people.
- Things possess and use energy.
- Energy is associated with movement.
- Energy is fluid and transfers when things happen.

There are more of these but this short list gives you an idea of how children discuss energy and its uses.

Teaching energy is a complex and difficult thing. Richard Feynmann, a Nobel prize-winning physicist and educator, wrote in some detail on the topic. He said that he did not know what energy was but that he could teach you how to calculate it in a great variety of ways! It is perhaps better to accept that we use the word 'energy' to explain what happens when substances interact. Rubbing our hands together to get them warm for example can be explained in simple terms, such as:

* We move our hands quickly.
* Friction between our moving hands generates heat.
* The movement energy has been transformed to heat energy.

Another example could be based on when a piece of wood burns. We could describe this by saying 'energy is released as heat to the surroundings'. We could then compare a piece of coal or wood say, and make an estimate of the energy 'released' by burning these. We don't really know what energy is but it helps us explain the phenomena in more detail. We could then go further and say that heat energy is given off when objects burn. The key is to match our explanations to the children's level of understanding whilst trying not to impose a rigid explanation that becomes almost rote.

Forms of energy

There is a wealth of literature in scientific circles as to how many forms of energy there are. It is not the intention of this text to delve deeply into this topic but to provide a simple and appropriate range of explanations and descriptions. Other chapters in the text will explore some of these in more detail to enable you to approach the topic with confidence.

Simple forms of energy which children can use to explain and explore the world around them include:

* light energy
* sound energy
* heat energy
* chemical energy
* stored energy
* movement energy
* electrical energy.

These are the most common forms in everyday use, however they can cause difficulties. Even young children may ask what the difference is between stored energy and chemical energy. Both of these could be deemed to be 'stored'; for example the energy in a fuel is stored but could also be described as chemical energy. Accept either and use this to discuss and talk about the topic. This discussion is key – encourage the children to explain simple situations from their own experience, but also attempt to use the scientific terms related to this topic. Use a diagram or an image as a discussion starter and then ask learners to note or discuss their ideas in the situations depicted.

Using energy

An initial approach could begin by examining everyday items at home and in class and asking learners what energy they associate with these: televisions use electrical energy but emit sound and light, mobile phones use electrical energy (batteries) but give out sound, cars use petrol but move, people use food in order to stay warm, play, talk etc. Emphasise the forms of energy and how these are essential for everyday life. Initial strategies would include the forms of energy being associated with an object. Toys, rattles, lights, musical instruments etc. all have various forms of energy associated with them. This identification of the form of energy is appropriate for younger children.

It leads on to describing items in terms of the energy required to operate these and the energy associated with its use: an electrical heater for example uses electrical energy and transforms it into heat energy; a music player uses electrical energy and produces sound energy; a remote-controlled car uses electrical energy (batteries) and produces movement energy.

This implies that in order for things to work there needs to be some form of energy 'in' and this leads to energy 'out'. This is essentially the conservation of energy. No energy is created or destroyed, it is just transformed from one form to another.

A strategy here is to initially ask the children to describe what was happening in the examples given. These could include:

- The sunlight heats up the pavements.
- The cars need fuel in order to move.
- The footballers need energy to play the game for 90 minutes.

Having established a basic description of what is happening, you can explore this further by asking 'How does this occur?' This has to be linked to the example the children have focused on, but during a discussion on heating and drying could lead to 'Why do the pavements become dry after rain?'

The sunlight heats up the pavements is a good descriptor, and as the teacher you could follow this with 'Why do the pavements become hot?' This could lead to *The sunlight (energy) is transformed to heat energy* or *The light energy is changed to heat energy*. This could be followed with 'What then happens to the water if the pavements become hotter?' This could then result in *The water becomes hotter and evaporates*.

This is obviously an ideal situation, but the methodology is to ask questions where the children's explanations are guided and shaped towards using energy terms to explain everyday events. Even if the children's responses are unstructured and limited, you should be encouraging them to ask why does this happen and suggest explanations.

Heat energy and temperature

Heat energy is associated with hot objects, hot days, fires etc. Friends and colleagues may discuss how to stay warm on a cold day by staying indoors or possibly 'wrapping up'. They may use phrases such as 'I need to heat up' or 'I hate the cold'. These phrases are common but can lead to misconceptions regarding the nature of heat energy.

During winter months people are encouraged to stay warm by reducing heat loss or conserving heat (energy). This often leads to 'keeping the heat in' or 'letting the cold in'. However there should be no mention of 'the cold' or cold energy in scientific terms. This is because there is no cold energy. It is the transfer of heat that makes objects hot or cold. 'Hot' and 'cold' are adjectives to describe temperature. Heat is a form of energy that can be transferred to make objects hotter or colder. Cold is *not* a form of energy and cold is not an antonym of heat.

To boil water we place the water in a pan and heat that pan by placing it on top of burning gas or a material that is at a high temperature. The hot material or burning flames transfer heat to the pan and from the pan to the water. The water increases its temperature – it gets hot. If we move the pan to a cooler area it will transfer heat to the area around it and cool down. Heat itself does not travel up nor down per se, it travels from a hot object to a cooler object. If we leave a hot cup of tea outside on a cold day, the tea will transfer heat to the colder air and cool down. When it reaches the same temperature as the air it will not be hotter or colder than the air, and therefore will not transfer any heat. It will remain at that temperature.

A hot room will transfer heat to the colder air outside due to heat energy passing through windows and walls. If we do not supply more heat to the room (keeping the radiators on for example), the temperature in the room will fall until eventually it will be at the same temperature as the air outside. Once it is at the same temperature no heat will be transferred and the room will remain at that temperature.

When we place our hands on a table our hands transfer heat to the table. We don't see or notice this easily, but where we have been in contact with the table it is hotter than the rest of the table. In a few minutes however the table will transfer this heat to the room and will eventually be at the original temperature.

> To see video clips of thermal camera imagery and using a thermal camera go to the companion website for this book at
> https://study.sagepub.com/chambersandsouter   ▶

Children have many misconceptions in this area. The word 'cold' is used in the opposite sense to heat. This is wrong. Hot and cold are opposites. Heat is a form of energy.

For example, when you go to bed at night does your bedcover keep the heat in or keep the cold out? The answer is it reduces the heat energy being lost from your body. *It does not keep the cold out*. Cold does not move. It is a description of the temperature of an object. The bedcover doesn't keep all the heat in. If your bed is hotter than the air temperature in your bedroom, heat will transfer from the hot object to the cooler object. A thick bedcover *reduces* the rate at which heat is lost from you, it doesn't keep it all in so to speak.

This reduction of heat loss is the same as in polystyrene coffee cups or double glazing or thick ski-type jackets. They reduce the rate at which heat is lost or transferred from a warm object.

This misconception is understandable. If a window is opened for example, *cold air* can come in and make a room cooler. This can be interpreted as the 'cold' coming in. Cold air enters the room and displaces the warm air to make the room cooler. The warmer air has moved around and out of the room.

Does heat rise?

As said before, heat travels from a hot 'object' to a 'cooler' object. However, if we hold our hands above a small fire or a candle we will feel more heat than if they were at the side. This is interpreted as heat rising.

In these cases hot air and hot water are rising. The air around the candle flame is heated and becomes less dense. This heated air rises and this is what we feel when we put our hands above the flame. The same is true of the water. It heats up, expands, becomes less dense, and rises to the surface. It then cools, falls to the bottom, is reheated, and then rises again in a convection cycle. Kettles have their elements at the bottom. The water heats up, rises, and is circulated.

A log fire for example has flames and ashes rising from the burning wood. The hot air from the fire rises and colder fresher air is drawn in. It is hotter above the flames than it is at the side. That is why we have chimneys. They allow the hot air to rise and leave the room and take all the dust and soot along with it.

In your home the temperature at the top of a room will be greater than that at the bottom. The hot air rises and the area close to the ceiling is generally warmer. The air temperature around the bottom of the room is colder. Many people like to wear slippers and thick socks to reduce the rate of heat loss from their feet (to keep their feet warm in normal conversation). This is because their feet are generally at the bottom of a room where the air is coldest.

To answer the original question 'Does heat rise?', the answer is 'not really': hot air rises, hot water rises, but heat travels from a hot object to a cold object.

How do radiators heat the room?

Radiators are generally filled with water. The water is heated in a boiler and then fed to the radiators throughout a building. Radiators can have fins or gaps between two metal plates. Heated air at the top of the radiator rises and cooler air is drawn in through the bottom. Radiators heat a room by *convection*. The hot air rises, moves about the room, cools and falls, and is then drawn into the radiator, is (re)heated and fed through the room again. Radiators don't radiate that much!

Energy sources and energy resources

These terms are often seen as synonymous. It is suggested here that for young children they are used as if they are the same thing. Many texts list sources and resources and there are common entries in both. A source of energy could be simply introduced as something that gives us energy or allows us to use it for our purposes. Coal is often described as a source of energy – we can utilise it to heat our homes and water, generate electrical energy, or extract various products from it to use at a later stage. The same is true of oil. Natural gas is mainly burned directly and this is used for heating and electrical power generation.

Using these terms in simple, straightforward explanations is crucial in our development of the concepts of energy for young people.

Photo 14.1

iStock.com/Jennifer Sharp

Children can be asked to describe what is taking place in the photo. Three widely varying and possible examples are given here:

> 'The coals are burning. As they burn heat energy is released. This heat energy is transferred to the food and its temperature is increased. This cooks the food. The sides of the barbeque get very hot and a lot of hot air rises from the coals heating the air in the garden'.

'The coals give out heat (energy). This cooks the food and makes it warm for us to eat'.

'The heat from the coals goes into the food and cooks it'.

These three possible descriptions show varying degrees of accuracy but none are essentially wrong in terms of a young person. You would try and expand the final answer by asking some questions which may guide the pupil to a more detailed explanation, such as:

- 'How does the heat get to the food?'
- 'What happens to the food?'
- 'How does the food get warm?'
- 'What happens if we leave it for too long next to the coals?'

This sort of approach should allow you to gain an insight into children's understanding of the topic and what needs to be done to remedy any misconceptions.

Batteries and electrical energy

Batteries are an obvious source of energy. They supply energy for simple circuits and allow us to light bulbs, run simple motors, use our mobile devices and so on. Our National Grid distributes energy from power stations up and down the country. Electrical energy production and distribution are critical to the smooth operation of the country.

It is the ease with which electrical energy can be distributed that makes it so important. In early societies populations grew up around rivers and ports or near supplies of wood or coal. Rivers allowed the easy transport of large amounts of materials, and a community built near to a forest for example would have wood for heating and such. Electrical energy can be generated in some distant location but this can be distributed very quickly to other locations. This allows communities to develop in remote places. The National Grid is the name we give to the series of electrical power cables and transformer sub-stations which distribute energy throughout the country. Many children are unaware of these. Their understanding of how we obtain energy from a plug socket could be very limited. The idea that when we plug in and switch on a kettle this creates a demand for energy, that has to be met by some coil of wires rotating inside a generator over 100 miles away, is not straightforward.

In a simple circuit we can operate a small bulb by connecting some leads to a switch, a battery and a bulb. By pressing the switch we complete the circuit and the bulb lights. In explaining and describing circuits like these the simplest explanations are the best: *When the switch is pressed the circuit is complete and the bulb lights. The energy travels from the battery to the bulb.*

This description is clear and to the point. If we start to make it more complex we lose meaning and can make the process much more difficult. For example: *Closing the switch completes the circuit and the electricity flows to the bulb making it light.*

This seems reasonable unless someone asks what electricity is. 'Electricity' certainly doesn't flow or move. Additionally, what is meant by the term electricity? Current? Energy? Power? Charge? These concepts are difficult enough. Keep it simple and describe what can be observed. The energy does indeed travel from the battery to the bulb. How it does so is incredibly complex. Some texts attempt to give a degree of understanding to the concept by relating the flow of energy to the current in a circuit. They often state or imply that the charges move round the circuit 'carrying' the energy to the appliance. Charges do indeed move around the circuit but do not carry the 'energy'. One needs to think carefully about explaining energy-related phenomena with moving charges. Keep explanations related to the energy supplier and the energy users. A complete circuit will transfer energy from the supply to the motor or light bulb.

The term power is also used widely in this context. Power stations generate energy and this is distributed around the country. Some electrical heaters are described as powerful. 'Power' refers to the rate at which energy is used or generated. A low-power bulb of 25W for example could still 'use' the same amount of energy as a high-power bulb of 40W. It would just have to be switched on for a longer period of time. A kettle rated at 2000W will not heat up as quickly as a kettle rated at 3000W. The unit of power is the watt, named after James Watt, a Scottish engineer. In energy terms, if you expend 100 Joules of energy in 5 seconds that is the equivalent of 20 Joules in 1 second which is 20W. Do not use the word 'power' when describing energy in a circuit (i.e. do not say 'The power comes from the battery' or 'The power goes to the motor'). Use the word 'energy' consistently. Using 'power' here leads to learners believing the terms are interchangeable. It therefore adds a level of confusion to an already complex situation.

Conservation of energy

This is the cornerstone of all discussions relating to energy. Energy cannot be 'lost' or appear to come from nowhere. It can be transformed into other types but the overall amount of energy remains unchanged. This can lead to complex discussions around the nature of energy and related topics. For example, the United Kingdom has a plan to reduce our use of fossil fuels. Some difficult questions could be posed regarding this. If you have been discussing energy conservation in that the energy must convert to other forms but not disappear, a reasonable question could be 'Why reduce?' If energy cannot be destroyed it is still there so we should be able to use it.

It is concentrated stores of energy that need to be conserved or saved (such as oil etc.). These are valuable as they have a high energy density (i.e. lots of energy per kilogram) and a few lumps for example can heat a room for hours, or generate millions of Joules per second in a large power station. Note that these sources are finite and will not last.

'Wasted' energy

In everyday life a lamp of 40W will keep a section of that room illuminated. Not all of the energy will be used as light – some of the energy will be transformed to heat. This is sometimes referred to as 'wasted energy'. It is wasted in that the energy is not being used directly for the intended purpose (see Figure 14.1).

Figure 14.1

The number 40 was chosen as 40W is 40 Joules/second. So every second this lamp produces 30 Joules of heat energy and 10 Joules of light energy. The light energy or the energy that the appliance was designed for is occasionally referred to as the 'useful' energy. You could ask 'what is the "useful" energy given out by a kettle?'. The answer is of course 'heat'.

If you ask the same question without the 'useful' term you may have to accept sound or light or even movement as kettles do 'waste' a portion.

This highlights the fact that much of our energy is used to do other things that we don't really want. Most electrical devices heat up and transform electrical energy to heat. In a TV or sound system we don't want this to occur. We want to watch TV, not use it to heat the room. This leads to the term energy efficiency. This is a measure of how much energy the device consumes and relates this to how much energy is given out as wasted heat. Many white goods such as fridge freezers and washing machines have an energy rating associated with them. This allows us to make a choice based not only on how an appliance looks but also on how environmentally friendly it is. The greater the number of energy efficient machines we use, the less energy is wasted. As a result there is less of a demand on our power generation network. This will help us reduce the rate at which we use up finite energy resources.

Summary

This chapter gave a good, working understanding of energy and associated issues relating to the teaching of the subject. It is conceptually very difficult and abstract, and further study can result in encountering some very obscure, almost philosophical ideas. Energy, its applications to real life and our sustainable future, however, are very important. A solid grounding in how we use, store, manage and transform our energy is the only way we can develop people who are informed and capable of addressing energy issues in the future.

REFLECTION POINTS

- How would you respond to a child who asks 'What do you think energy means?'

  - A difficult question! You could begin with the idea that we need energy in order for things to happen or to work. We need food to allow us to grow and survive. We need energy so that our factories and industries can function. Animals and plants need energy in order to grow and breed. Our household devices need energy so that we can watch TV, stay warm, cook food etc. If we didn't have energy originally nothing would have started and we need energy to continue our existence. We use energy to explain how things work and it helps us understand what we need to thrive.

- How would you explain how ice cubes can cool your drink?

  - Placing ice cubes in your drink will chill it. Ice is cold, less than 0°C. Your drink will be around 15°C. Heat will flow from a hot object to a colder object so heat will leave the drink and flow to the ice. It takes a large amount of energy to melt ice so all this heat energy flows from the drink to the ice, thereby cooling the drink.

- Where does the electrical energy we use to operate our lights come from?

  - In power stations concentrated sources of energy, such as coal, oil or gas, are burned in a boiler. This heats large amounts of water until it is converted to steam. The steam turns a turbine which generates electrical energy. That electrical energy is distributed via the National Grid to homes across the country. When we switch on our lights the electrical energy is transformed to light energy and we can see to move around our homes.

15.
ELECTRICITY

electric

CHARGE

ON
OFF

Learning Objectives

Having read this chapter you should be able to:

- know that all objects are composed of charged particles
- be able to explain simple electrostatic phenomena in terms of moving charges
- be able to explain why people may receive tiny electrical shocks under certain circumstances
- be able to describe and explain simple electrical circuits using terms such as insulators, conductors, voltage and current

Overview

Electricity is a general or all-encompassing term. It is used to describe a range of phenomena – from objects being rubbed and charged, to how we transfer energy, to the basic principles that operate the technological devices we rely upon today.

In order to understand and explain the basic principles involved in electrical phenomena we need a basic understanding of the structure of the atom. An atom consists of a central portion – the nucleus – which has a positive charge. Electrons orbit this nucleus and are negatively charged. Charge is a basic and fundamental property of matter and two 'types' have been identified: 'positive' and 'negative'.

Electrons have a negative charge and the nucleus has a positive charge.

In normal circumstances the total numbers of positive charges and negative charges are balanced: the negative and positive charges 'cancel' each other out to give an overall charge of zero. In this situation the material is said to be neutral (i.e. having no overall charge).

The ability of certain materials to attract tiny objects has been known for a very long time. Rubbed amber was known to attract small objects. The Greek word for amber is *electron* and this is the basis for the word 'electricity'. The amber could be rubbed enough to cause small sparks.

Many materials become charged when rubbed against other materials. Rubbing a balloon against your hair or shirt can charge the balloon. This balloon can be used to pick up small pieces of foil and tissues. Rubbing the balloon in your hair for example can attract your hair, and you can hold the balloon above your head with your hair attracted to it.

These experiments are very suitable for young children and work well. They can be used to determine which materials work well and which do not.

In general most insulating materials such as plastics can be rubbed and become charged. Plastic pens, hairbrushes, disposable cups and rulers for example all work well. Note that metal objects do not.

Explaining the experiments

Your explanation of the events should focus on what the children have done. Emphasise that a rubbed object becomes 'charged' and that a charged object and an uncharged object will attract:

* The charged balloon attracts the small paper and foil.
* The charged balloon is attracted to the wall and 'sticks' to the wall.
* The rubbed straw has become charged and is attracted to the finger.
* The charged balloon attracts the water and is deflected towards it.

Many children and students may use phrases such as 'opposites attract' which is correct but this can cause confusion. A difficulty with this is that the paper, foil and wall are not obviously charged, they are neutral. To ensure a consistent explanation we would suggest that you state a *charged* object can attract an *uncharged* object. Explaining these phenomena in greater detail can mean encountering concepts beyond the scope of this text.

It is worthwhile to consider the phrases used in these experiments. Many will ascribe the word 'static' to these phenomena. It explains nothing and is used as a 'catch all' phrase. 'Static' means not moving. When we rub a section or certain area of a balloon, it is that part that becomes charged and remains charged. The charges do not travel across the balloon or hairbrush, they remain in that area. They are 'static'.

How do objects become charged?

When two materials with the appropriate properties come into contact electrons can be transferred from one material to the other. The material that gains more electrons gains more negative charges. It becomes *negatively* charged.

The material that loses the electrons has an imbalance of charge. It now has more positive charges than negative charges. It is *positively* charged.

No new charges were created in this process. Existing charges have been separated. This process is essentially *triboelectric*. 'Tribo' means rubbing or friction. When the two materials are brought into contact with each other, charges are transferred. The act of rubbing increases the number of contacts and allows more charge to be transferred.

This idea that we can charge certain objects by rubbing them may be difficult for young children, and should be introduced by allowing them to rub various materials, pens, pencils, rulers, rubbers, paper and the like, and then try to lift or raise small objects like little scraps of paper tissue.

Electrostatics experiments work best on days with low humidity (e.g. a good summer day or a nice frosty morning). If the weather is humid, dry the room, turn the heating up, and if necessary run a hairdryer for 15 minutes or so before undertaking experiments.

The experiments discussed earlier should involve rods placed on watch glasses or hung from thread. This is so that any slight force acting on them will cause them to move and this can be seen by the pupils. If we bring a charged object close and one person is holding the rod, they cannot detect the force as it is too small. If the rod is on a desk it won't move as the force pushing it is not large enough to overcome the friction between the rod and the desk. The watch glasses and thread are the best ways of observing the effect.

Objects with the same charge

Charge two identical pieces in the same way (e.g. rub both with the same duster). Place one on a watch glass and bring the other to it. You will find that they will push each other away – they repel each other. Similarly charged objects repel (push away). When people become charged and some of these charges are transferred to their hair, their hair will be repelled from their body. This will cause their hair to stand up or away from their head. The charges are repelling.

Sparks and discharging

If a plastic rod is rubbed, or a person walks across a carpet, or that person sits on a car seat for a while, they can become charged. This excess of charge will seek to balance itself (i.e. it will move to another material to make your overall charge neutral, or other charges will be drawn to you to make you neutral). This is most easily done when you touch a conducting material like a door handle or even a car (or a person). Often these sparks will give the person a fright and can be painful at times. Many people are reluctant to close car doors for example for fear of being shocked. In most cases the effects are minimal, but there are certain occupations where this sparking could be dangerous – working with fine powders or flammable materials for example. Workers in these positions will often have to earth themselves prior to working.

Everyday examples of static charge build-up

Children on plastic chutes and trampolines

As they slide down or jump, their clothes rub against the plastic and they become charged.

Repeated slides and jumps mean they can become quite highly charged.

If they become charged and they have fine hair which becomes charged their hair can stick out!!

This is explained by the pupil and their hair having the same charge. The same charge repels and pushes away so their hair can be seen to stick out.

Supermarket trolleys

Many supermarket trolleys have nylon wheels which rub against the plastic tiles as they are pushed around. These wheels become charged and can then transfer this charge to the trolley. A child in a charged trolley can often be seen to have their hair stick out. If the person pushing the trolley touches it, it can discharge, and the person will feel a spark or a small shock.

Getting out of cars

As we drive or sit in a car we move about and brush against the seat and our clothes rub against each other. This generates charge and as the car acts in a certain way (due to it being like a Faraday cage which is a complete shell of metal which allows charge to pass almost completely around the shell) we can get out of the car and be quite charged. When we then touch the car door to close it, we can discharge it and get a shock.

Lightning

Lightning is generally caused by bodies of air rubbing against each other and becoming charged. Rising hot air comes into contact with falling colder air. The air becomes charged and the lightning discharges this excess charge.

This is generally true and should be used in your explaining lightning to learners.

The precise mechanisms of how and when lightning occurs are complex and far beyond the scope of this text.

Questions and answers relating to static electricity

Why do we get shocks?

Charge can often build up on people and reach levels that will give uncomfortable shocks, damage sensitive electronic parts, or create fire risks when handling solvents and other flammable materials. Dry, arid days with low humidity encourage static charge build-up and electrostatic discharge can be strong in these conditions. In the UK the worst case is often in winter (January to March), when cold and dry external air is warmed and brought into a building, leading to very dry internal atmospheric conditions.

There is a wide range of factors that can affect the amount of electrostatic charge that can build up as a voltage on people. Some common major factors include:

- the floor material and its electrical properties
- personal footwear, especially the materials making up the soles of shoes
- atmospheric humidity
- the manner in which a person walks (e.g. scuffing and friction of shoes against the floor)
- the action of brushing against furniture, sitting and rising from seats.

Why do we get shocks when touching a door knob or filing cabinet?

Most modern shoes have highly insulating rubber or plastic soles. As we walk, static charges build up on the soles of our shoes. This is especially true if the floor is also insulating. Some older nylon carpets are particularly good at generating static charge.

The charge on the shoe soles passes to our body, and this charge can generate a high voltage. Under severe conditions more than 15,000 Volts have been recorded. It is quite common to experience 5000 Volts. In fact, many people do not feel a shock from a static electricity discharge less than about 2000 Volts.

I get shocks when I'm sitting, or getting up from a chair – and I haven't walked anywhere!

When you sit in a chair the contact between your clothes and the chair can generate a lot of electrostatic charge. If you lean forward so your back moves away from the chair back, or if you get up out of the chair, the charges separate and you can take the electrostatic charge with you. Your body voltage can rise very rapidly as the charge is separated from its charge on the chair. Furniture covering is a major factor here.

Why do I experience shocks when my family and friends don't?

There are many reasons why this might happen. Firstly, some people are more sensitive to shocks than others. Secondly, you may be storing more electrical charges than others. This depends on the size of your body and feet, even the thickness of your shoe soles! A bigger body, bigger feet, and thinner shoe soles, mean more charge has to be stored to produce the same voltage. Thirdly, you may be generating more charge than others. This may be due to the material of your shoe soles, or the way that you walk. If it happens when sitting, it may be due to the material of your clothes, and the amount of static they generate against your chair.

Shocks when using a vacuum cleaner

When dust travels in the air sucked through a vacuum cleaner, it rubs against the pipe walls and other internal parts. This rubbing generates separate charges on the particles and on the pipe walls. This can lead to shocks when touching parts of the cleaner.

Why don't I get static shocks when I touch something like a wall or a tree or a door?

If the wall or door is made of wood, concrete, or some other material that has low conductivity, any static charge on your body escapes slowly and usually does not cause a shock. If you touch metal, water, or another person when your body is highly charged, the charge is discharged quickly as these materials are highly conductive. In this case you may feel a shock.

I get shocks when I'm shopping – how can I stop this?

Once again, you build up electrostatic charge as you walk around. However, if you're pushing a trolley, the wheels of the trolley can also generate electric charge. As you walk around, you and the trolley both store charge and reach a high voltage. When you reach to touch something, you get a shock. To reduce the chance of receiving a shock you should touch metal objects frequently to discharge you before it builds up to a high level.

What about shocks when I get out of my car?

Many people experience shocks when they get out of their car. Often they will believe that the car is charged – but this is not normally so. Sitting in the car, electrostatic charges are generated on the car seat and the person's body, due to contact and movement between the clothes and the seat. When the person leaves the seat, they take charge with them. As they get out of the vehicle, their body voltage rises due to this charge. When they reach to touch the vehicle door, the electrostatic discharge and shock occur as their hand approaches the metal door.

Electrical circuits

Our modern lifestyle is energy intensive. We all use devices which are reliant upon electrical energy. This energy is generated and then distributed to our homes and work. Without it our laptops, phones, tablets, cookers, lights and such would not operate.

In the previous section incidences of static charging were looked at. Materials were described as being charged due to an imbalance of the charges on them. The charges could not move easily, they were 'static'. This section looks at the movement of charges in a wire. The movement of charges in a wire is referred to as an electrical current. In order for a simple circuit to be constructed we need a number of things. We need a battery or source of energy and a series of connecting wires. This forms a circuit. There are many educational electrical resources available and the level of experimentation

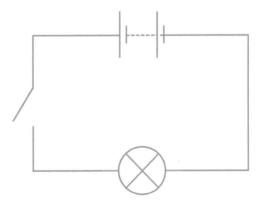

Figure 15.1

will obviously depend upon the resources at your disposal. The process of allowing children to build, test and fix (fault find) their circuits is fundamental to working with circuits. To attempt to explain every process or circuit they build is challenging. Setting small problems where they report to you or light up bulbs is a good outcome.

In this circuit we have a source of energy (a battery), some connecting wires and a lamp. The energy from the battery travels to the lamp and the lamp lights up. We need a series of conducting wires to allow a current to flow and this current allows the energy to flow to the lamp. When answering questions regarding these sorts of circuits, the key is to look for a complete circuit of conductors and an energy supply.

Simple circuit symbols

It is sometimes difficult to draw or describe these circuits in 'longhand' terms so it is worthwhile introducing simple circuit symbols. These make the circuits straightforward to follow, and it is a good exercise for children to compare the actual circuit with the 'symbolic' circuit.

Figure 15.2

Comparing circuits

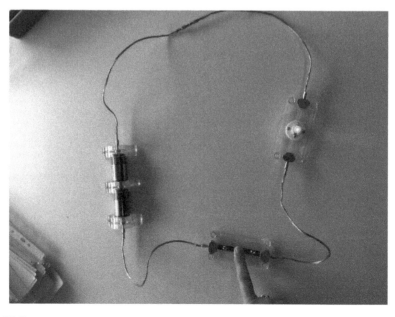

Figure 15.3

The two circuits shown in Figure 15.1 and 15.3 are identical. The circuit diagram in 15.1 is a representation of the circuit in 15.3. When the switch is pressed the circuit is complete, a current can flow, the energy is transferred to the bulb, and it lights.

The supply voltage

The supply voltage is a measure of the electrical energy given to a circuit. Most batteries have a rating of 1.5V. If two batteries are connected together in series (side by side) the supply voltage will be increased to 3.0V. Three batteries would give 4.5V and so on. The greater the number of batteries in a circuit, the greater the amount of energy available to that circuit.

To see video clips of electrical circuits and charged objects go to the companion website at https://study.sagepub.com/chambersandsouter

In general, increasing the supply voltage increases the volume of your music player or the brightness of a lamp. If the voltage is increased the filament in the lamp can get too hot and melt. Adding more bulbs to the circuits will result in the bulbs getting dimmer. The supply voltage is shared between the bulbs and as a result the bulbs are not as bright. This demonstrates how the 'voltage' in a simple series circuit is shared between the components.

Current

Most conductors are metals. The atomic structure of metals is such that it is relatively easy for charged electrons to flow through them. This flow of charge is referred to as the current and is measured in Amps. The current is directly related to the amount of charge flowing in the wire.

In the circuit shown, when the light is on a current is flowing in the circuit. The current at point A is the same magnitude (size) as the current at point B and point C.

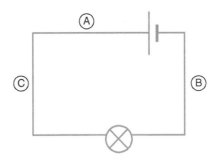

Figure 15.4

This is because the current is related to the amount of charge flowing in the wire. There are no other routes for the charge to flow so the same amount of charge flows in all parts of the circuit. A current of 0.3A for example will be the current at all points in the circuit.

Conductors and insulators

As stated earlier, metals are generally good conductors. They will allow a current to flow in a circuit and the circuit will light a bulb, run a motor, or heat a wire. Other materials are not good at allowing current to flow. They are insulators. Their atomic structure makes it difficult for electrons to flow, and if the electrons cannot flow no energy can be transferred from the battery to the appliance. A simple but effective experiment to check for this is to build a circuit as shown.

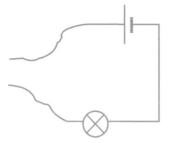

Figure 15.5

Place a range of everyday objects in the gap as shown. These should include paper clips, elastic bands, pens, pencils, rulers, coins, rings etc. The materials that allow the lamp to light are conductors and those that do not are insulators. A small investigation here is to ask the children to organise these materials into two groups. This will promote classification based upon conductive properties.

Some materials are better than others at allowing current to flow and they are said to have a low resistance. Copper is used for the wiring in homes as it is relatively inexpensive and has good conducting properties. Other metals are better conductors, such as gold, but the cost of wiring your house in gold is prohibitive. Some insulators are more effective at 'stopping' electrical current than others and these are used to cover conducting wires. Plastic and rubber are very good insulators and are used to cover plugs and cables. Plastic and rubber however are not good enough insulators for very high voltages and other materials are used.

Ceramic materials are utilised to insulate high-voltage power cables. These are used to hang the cables from the metal pylons which support them.

Photo 15.1

iStock.com/Robert Dant

They provide a very strong insulating barrier between the high voltage cables and the pylons. Under certain conditions even these insulators have been known to fail and arcing can occur.

Types of circuits

There are two main types of simple circuits that can be examined in the primary sector: series and parallel. Most circuits, even complex ones, are combinations of these.

Series circuit

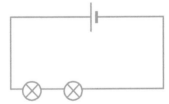

Figure 15.6

This is a straightforward circuit. All the components are connected to each other in a simple loop. The circuit is a simple series circuit with a battery and two lamps 'connected in series'. The lamps will light as there is an energy supply and a complete circuit of conductors. When there is a break in the circuit or if one bulb 'blows', all the other components will stop working. This is because there is no longer a complete circuit of conductors. This was once a problem with Christmas tree lights when if one 'blew' all of the bulbs had to be tested to determine which one needed to be replaced. Modern Christmas tree lights are connected in series but have a small wire across the terminals. In normal use it does not affect the circuit but if the filament breaks this wire conducts and allows the other bulbs to light. You need to replace these broken bulbs relatively quickly however, as the other bulbs receive more of the voltage and may ultimately overheat and fail.

Parallel circuit

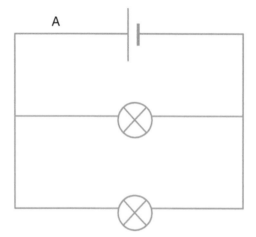

Figure 15.7

A parallel circuit is one where there are sections which are (unsurprisingly) connected 'in parallel'. The circuit shown has two lamps connected in parallel. The advantage of this is that each of the lamps can operate individually. If one lamp is off the other is not affected. This is because each section can be treated individually, as each has a complete circuit of conductors connected to a power supply and can work independently. Lighting circuits at home are wired in a type of parallel circuit which allows each light to be switched on and off independently.

In a parallel circuit the voltage across each bulb is the same as that of the battery. It is not shared or split between them. This is because each 'branch' acts as an independent circuit. As a result it receives all the voltage from the supply.

The current from the battery however must operate both bulbs. As said earlier the current is related to the actual number of charges flowing in the wire. Therefore the current from the battery at position A must split into two smaller currents through the bulbs. A current of 1.8A could 'separate' to 1.0A through one bulb and 0.8A through the other bulb. The current would then recombine in the wire just before the other terminal of the battery.

A fuller understanding of how electrical circuits function is very complex. Many texts use analogies in explaining how electrons move round the circuit, carrying energy and transferring energy to bulbs and motors. These can be problematic. These analogies can have some difficult stumbling blocks and obvious flaws. For example, in some texts current is often analogous to water flowing in a pipe. When we break a circuit (or the pipe) the current does not flow, but if it were a pipe the water would still flow. Electrons also do not move quickly through the wires and nor do they carry energy. Some explanations that appear 'wrong' may be perfectly acceptable as appropriate to assist in a child's understanding at that point. Allowing children to experiment with, design and describe how their electrical circuits work is recommended, as this encourages inquiry and creativity in this field.

Our national electrical supply

All the electrical energy we use in Great Britain has to be generated in some way. Power stations turn a turbine and this rotates wires within a strong magnetic field. This then generates the electrical energy which is distributed throughout the country by a series of pylons and wires. The grid is also connected to other countries via undersea interconnections. There are various sub-stations throughout the country which regulate the supply to ensure we have a steady 230V in our homes. Our modern appliances are designed to operate on a 230V a.c. (alternating current) supply.

The advantage of using electricity as our main source of energy for our homes is its ability to be sent from generation to consumer relatively easily. This allowed industries and cities to be established without the need for local energy supplies. Electrical power distribution started in the late 1800s and there were a number of competing systems. Distance was a problem as the electrical energy was dissipated in the wire if the stations were far from the consumer. High voltage a.c. was a solution to the energy being lost in the grid, and it also allowed other stations to join and be connected to it. Our distributor of electrical energy, the National Grid, takes energy from all forms of power stations in the country and distributes this to everyone in the country. It is an enormously complex and interconnected system but we very often take its existence for granted.

Electrical safety

Electrical circuits in school are very safe and there is little risk. Working with batteries and bulbs is almost risk free. Occasionally a bulb may heat up but otherwise it is low risk. We should however be aware of the dangers of electrical circuits at home. Our supply voltage at home is 230V. This can cause severe injury and death. Our batteries are rated 1.5V so will not cause any harm. Even putting three together will not injure anyone. Wiring in the UK has three wires, normally 'live', 'neutral' and 'earth'. If someone was to make contact with the live wire they would receive a severe electrical shock. The earth and neutral wires are required for the appliances to operate and for safety purposes. Today all plugs are moulded which does not allow people to re-wire a broken plug, only to replace the fuse essentially. Most homes have circuit breakers which will break the circuit if there is a change which could be caused by someone being in contact with a wire. Touching the mains is dangerous and a number of things can occur. Certain muscles can go into spasm and not operate correctly (heart, diaphragm). If there is short-term contact there can be burns where we touch the wire as well as where the current exits our body. The effects are difficult to estimate as much depends upon our fitness, age, size etc. If we are wet the effects are more severe as this reduces our resistance and increases possible risks. Thankfully instances of accidental electrocution are rare, but the risks should be explained.

Summary

This chapter provided an introduction to the topic of electrical phenomena. It outlined a relatively straightforward way you could introduce basic electrical terms such as charge and energy, and then progressed to cover more advanced concepts such as the movement of charge and basic electrical circuits. It also provided you with the answers to some possible questions or inquiries from children regarding sparks and minor shocks. This was done to give you sufficient confidence to be able to teach these topics without fearing you were reaching the limits of your knowledge.

REFLECTION POINTS

- A person rubs a balloon and places it on a wall where it stays. How would you explain this phenomenon to a pupil?

  ○ When the balloon is rubbed the section that is rubbed becomes charged. A charged object will attract an uncharged object. The balloon is brought close to the wall and is attracted to the wall. The wall cannot move. The balloon therefore appears to stick to the wall.

- A child builds a series circuit containing two lamps, a battery and a switch. When the switch is pressed only one bulb lights. The child asks why. How would you explain this?

- The fact that one lamp is lit confirms that the circuit is built correctly. There must be a complete circuit in order for one of the lamps to light. The fact the other lamp is not lit does not mean it is broken, it means this lamp needs a greater voltage to light. It is merely a differently-rated lamp. If you were to add another battery to the circuit there may be enough voltage for the second bulb to light. (This is a good circuit to explore as many children will say that one is lit because it is nearer to the battery. Allow them to rebuild the circuit with the lamps reversed and they will find the same lamp is lit. Note that the position of a bulb relative to a battery has no effect.)

- Are the lights in your home wired in series or parallel? Justify your answer.

  - The lights are wired in parallel. If one light was to break, all the other lights in the house remain working. Therefore it must be wired in a type of parallel circuit. The same is true of our plug sockets. If a socket is broken or disconnected the other sockets in our house still operate.
  - Occasionally a fault with a socket or bulb will lead to the rest of the sockets in one area of the house going off. This is not a fault of the type of circuit. The fault may have caused a sudden change in the voltage and caused the circuit breaker to switch off. This switches off all the circuits that are linked to the circuit breaker (e.g. downstairs lights or kitchen sockets). The circuit breaker has to be re-set and all the sockets will then be operational.

16. LIGHT

Learning Objectives

After reading this chapter you will be able to:

- know that light is a form of energy and travels in straight lines
- be able to explain why we see and recognise objects in terms of light reflection
- have an understanding of the structure and function of the eye
- be aware of the dangers associated in observing the Sun
- be able to explain shadow formation

Overview

Light is a form or type of energy that can be detected by our eyes. The term generally refers to visible light (as opposed to U-V light for example). Light is a form of energy, and whilst it is one of the easiest forms for children to identify with, it could more correctly be classified as a form of radiant energy. A light bulb for example will emit light but also some heat energy. Both of these could be classified as radiant energies but with slightly different properties.

For the purpose of this text we will refer to light as energy that can be seen by the human eye.

Our main source of light on Earth is the Sun. Sunlight is the main source of energy that green plants use to create sugars and starch. This process (photosynthesis) is the main source of energy for almost all living things.

How we see things

We see objects because light is shone upon an object and some of that light energy is absorbed and some is reflected from the object. This reflected light is generally reflected in all directions. The reflected light reaches our eyes; we recognise the size and shape of the object and can then identify the object.

Our eyes are very sensitive, complex and fragile organs. They 'collect' light from our environment, limit the amount or intensity of that light, focus it onto a sensitive section, and then convert it into a series of electrical signals which are sent to our brain.

As our eyes are very fragile our body has adapted to protect them. They are slightly recessed in our skull and this offers a degree of protection against a flat object impacting on our face.

Our eyelashes act as a line of defence. If any small object such as a fly or piece of dust strikes our eyelashes we blink. This is a reflex action and happens very quickly.

Our eyelids also act as a small barrier and close over the eyes. Additionally they also spread our tears evenly over the eyes to clean them. The moisture then drains into a small opening and from there to our nose.

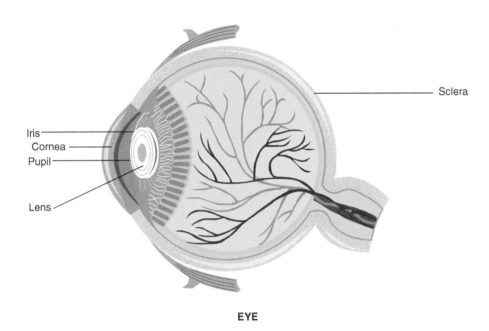

EYE

Figure 16.1

The front section of the eye contains the parts we can see in a mirror. This includes:

- the coloured part of the eye (the iris);
- the dark, circular part of the eye (the pupil);
- the white part of the eye (the sclera);
- a clear, curved section in front of the pupil (the cornea).

The eye operates by 'collecting' light which has reflected off objects. This light passes through the cornea and the lens to focus on the retina. This sends the electrical signals to the brain.

Figure 16.2 shows how the light is gathered and focused on the retina. A key point here is that the eyes take in light; they do not send out some sort of light rays. When we ask someone to 'focus' on a distant object, it can give the impression we are doing

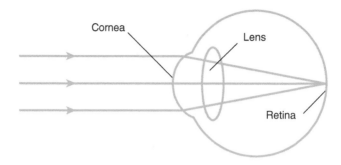

Figure 16.2

something external when we are merely looking at that object and focusing the light from it that enters our eyes.

How our eyes work

As shown above, the light from an external object reaches our eyes. Our cornea and lens manipulate (refract) the light to focus on the retina.

ACTIVITY

Making a pinhole camera

Collect the following:

A small cardboard box (shoeboxes are good).

A small piece of tinfoil (about 75mm x 75mm).

A sheet of tracing paper.

Sellotape, needle, scissors.

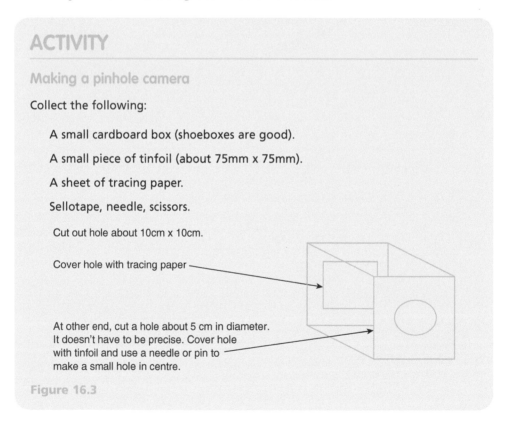

Cut out hole about 10cm x 10cm.

Cover hole with tracing paper

At other end, cut a hole about 5 cm in diameter.
It doesn't have to be precise. Cover hole
with tinfoil and use a needle or pin to
make a small hole in centre.

Figure 16.3

Ensure the shoebox is in good condition and that there are no openings that can let any light in once the lid is on. Taping any opening(s) closed with some tape will fix this.

Point your camera towards something bright (a bright lamp, a view from a window) and you should see an image of this on the tracing paper.

The image will be inverted and the children will say it is upside down which is acceptable at their stage. You can let more light in by making the hole larger but this reduces the sharpness of the image and it becomes slightly fuzzy. This is where a lens is useful. It takes the light from the wider gap and focuses it to produce a brighter and sharper image. The size of the hole corresponds to our pupil closing and opening.

This pinhole camera replicates the workings of the eye and can lead to some interesting teaching points. The image on our retina and the tracing paper is inverted but our brain rectifies this so we see what we do. Allow learners to suggest what would happen if the camera was to be turned upside down. This is an interesting and fun project with many applications.

To see videos on how to make a pinhole camera go to the companion website for the book at https://study.sagepub.com/chambersandsouter

In order for us to see and observe anything, the light must travel from an object to our eyes. As said earlier the Sun radiates most of the light we have on Earth. It also emits 'light' of other frequencies which our eyes cannot detect, such as Infra Red and Ultra Violet. The Sun is a radiant object. It generates and radiates this

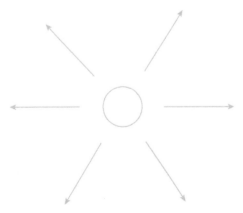

Figure 16.4

energy. To 'radiate' can be taken as to emit or give out energy generally in the form of rays. A good way to think of a radiant object is like a point emitting energy in straight lines.

Photo 16.1

iStock.com/hadzi3

Most of the objects we see in the sky at night are radiant. All the stars that make up the constellations and galaxies are undergoing nuclear reactions and radiating large amounts of energy. This energy radiates throughout the universe and reaches us.

The Moon and the local planets are bright objects in the sky but are not radiating energy of their own accord. They are reflecting the energy from the Sun. We can only see them because they absorb some light from the Sun and reflect the rest. Occasionally we will see some very bright objects in the sky but these can be the International Space Station or occasionally Low Earth Orbit satellites. We see these because the light from the Sun strikes them and is reflected to our eyes.

Observing the Sun

It is *not* good practice to look directly at the Sun. If you do so it can lead to your vision being damaged. If we catch sight of the Sun and it dazzles us we will generally turn our head and close our eyes. This is our natural response. If people do try to look for longer than a few seconds too much U-V can reach the retina, causing temporary blurriness or blindness. It is however painful and no one can really look for more than a second or two. Every time there is a solar eclipse hospitals will report a surge in eye problems as people try to look at the eclipse. If somehow you do manage to keep looking at the Sun, light sensitive cells will become overstimulated and release a batch of chemicals which will damage the eye. These effects tend to be temporary but it can take months for your eyesight to return to normal. Sir Isaac Newton damaged his eyes whilst researching the light from the Sun. Those living in very sunny environments for many years can develop eye problems. The best prevention for this is to wear glasses that block U-V light.

Lighting to live

We need to see in order to go about our everyday life so one of the earliest 'discoveries' by early humans was fire. In addition to the fire keeping us warm and cooking our food it allowed us to see and function. A large wooden fire is not all that transportable so we developed portable hand-carried 'flames'. Oil lamps would be an early example of this and these have been found to date from c.4000 BC. These led to candles and 'lamps' made of cloths or wicks soaked in animal fat. Roman societies did not have streetlights but the pavements were made with reflective stones embedded in them, which then allowed people to make their way home as the light from the Moon was reflected off those stones.

House and street lighting

William Murdoch (Murdock), a Scottish engineer who spent almost all of his working life in England, is credited with using the gas given out by heated coal as a form of lighting. This led to some small gas streetlights and his home being lit by gas in or around 1792.

Our homes and streets are generally lit by electrical lamps. These provide reasonable vision and allow us to go about our business. Cars use electrical bulbs to let us see them from a distance and allow drivers to see the road ahead. These are all *radiant* objects: they emit light, which is reflected off objects; this light is reflected to our eyes and we interpret what we see.

How 'good' are our eyes?

Our eyes gather the light entering via the pupil and focus it on our retina. However we cannot see distant or small objects as easily as close or large objects. Some interesting experiments can be done to examine what our children can see.

ACTIVITY

Field of view

This looks at how widely we can see without having to move our heads from side to side. It can be done as a pupil activity. Ask a pupil to sit on a chair in the classroom. You or other pupils will walk around that pupil (in an arc) holding a card. The pupil looks straight ahead and indicates when they cannot see the card. Repeat this experiment for a number of pupils and record/discuss the results. In general, we can see a very large field of view without the need to move our heads.

Visual acuity

How well can we detect small or fine details? The term 20/20 vision is related to this. It was based on a standard-sized image that can be seen accurately from a distance of 20 feet. The letters seen in an eye test were originally designed to be viewed from a distance of 20 feet. If you could see this object at a distance of 20 feet you would have 20/20 vision. If you could see a detailed image of an object that a 'normal' person could see at 50 feet you would have 20/50 vision. In European and Australian regions this is sometimes written as 6/6 vision, where the '6' refers to the distance in metres.

ACTIVITY

Visual acuity

Children can perform little activities like this by printing off a series of letters and then getting their partners to move further and further away until they begin to lose the detail. Some children will be able to see further/closer than others, but in young people you can inform them that as they are growing the size and shape of their eyes will also change over the years, and that the small tests just now are only to discuss how they see and not to make any measurement of their vision. This allows for some interesting scientific procedures and measurement.

Colour vision

Most of us are aware that white light can be separated into various colours. Sir Isaac Newton showed this by passing white light through a glass prism. The colours observed range from red at one end to violet at the other. Many texts give red, orange, yellow, green, blue, indigo and violet as the colours of the spectrum. We perceive colours when light strikes the cone cells in our retina. Humans have three types of cones and these are responsible for our colour vision.

The physics of colour is complex and the perception of colour by different people is a subjective psychological phenomenon. What one person sees as a particular shade or colour may not be the same for other people. Most people will accept or agree to general colour similarities and identify these as the same colour or shade, although there will be occasional differences.

Very few (if any) people are colour blind. Most people who struggle to identify colours are colour deficient. This is when the cones in their retina do not fully respond to the light that strikes it and they perceive it as a different colour or shade. In cases like this reds and greens can be confused or a person will be unable to see these as different. The person still sees correctly but perhaps not the range of colours of other people with normal colour vision. Most colour deficiency problems are hereditary, and males are far more likely to be colour deficient than females. The gene that causes this is X-linked recessive and passed via the mother.

Ishihara colour tests are widely available online and it is worthwhile to allow the class to undertake these tests and then compare if any of the children 'see' the numbers differently.

How light travels

As can be seen from the earlier diagrams light travels in straight lines. In order to see an object we will generally need a line of sight (i.e. a straight line between the object and ourselves).

A torch or a simple laser pointer can show light travelling from a source to a screen or reflective object. Most children would accept that the torch shines the light onto the wall and that the light travels from the torch to the wall. The light from the torch spreads out as it travels and as a result it gets dimmer or less bright the further away it is seen or shone on a wall. The light cannot be seen as it travels from the torch to the wall. It can only be seen when it strikes an object and the light is reflected into our eyes.

Photo 16.2

iStock.com/Jiripravda

In the photo this sort of image is incorrect. The picture has been adjusted to show the path of the light but this would not be seen under normal conditions. We can exemplify (and amaze) children using a laser pointer and some fine spray from water, antiperspirant, or even chalk dust.

Place the pointer on a desk and point it at a board or a screen across the room. (Ensure that the pointer is pointing away from the class and they have no access to it.) Learners will see a red dot on the wall. Spray a fine mist or chalk dust over the air along the path of the light. They will then see the beam appear along the line from the pointer to the board. If the pointer is moved the line moves in accordance, and it is obvious to see the light travelling from pointer to screen.

This experiment may be difficult with primary classes but it is replicated here.

Light travels in straight lines but can be made to alter and change the direction it is travelling. We can shine light at an object, but if a shiny/mirrored object is placed in its line the light will reflect off that object and travel in another direction. This property of light (reflection) allows us to devise artefacts that allow us to see things we would not normally be able to.

A mirror is a piece of glass which has had a silver coating applied to the far side of the glass. This allows the light rays from a person to pass through the glass to the silvered side which reflects most of these back. It allows you to look at your image to check that your face hasn't changed dramatically since the previous evening!

Mirrors reflect images back at the same angle.

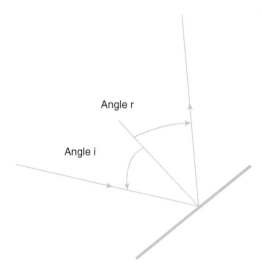

Angle r

Angle i

Figure 16.5

In Figure 16.5 the angles i and r are the same, and if we change the angle i the light strikes the mirror and the angle r changes correspondingly. This is known as the law of reflection.

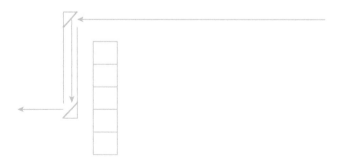

Figure 16.6

This principle allows us to see objects that could be above a wall or around a corner for example. Using a square cardboard tube and two mirrors we can make a simple periscope which will allow us to do these things. They operate on the principle of light reflecting off the mirrors into the eyes of the observer.

In Figure 16.6 the light from an object on the other side of the wall reaches the top mirror, is reflected down to the bottom mirror, and reflected into the eyes of the observer.

Some country roads which are obscured by hedges have mirrors at the entrance to allow drivers to see if there is additional traffic before they join that road.

Parabolic mirrors

A bulb will radiate light in all directions. This is of use when the bulb is in the corner of a room or on the ceiling and the light is radiated and reflected in all directions.

If we want to light up a particular area for example this is not very effective as much of the light is radiated in other directions. A possible solution to this would be to add a mirror to the back of the bulb and this should reflect the light in the direction we want.

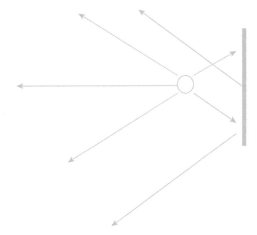

Figure 16.7

This combination reflects more light in one direction but it still does not allow us to point it or aim it with any great accuracy. We can improve this by bending or curving the mirror behind the lamp.

This is called a parabolic mirror or parabolic reflector.

This directs the light in the direction we want and is used in torches, streetlights, spotlights, car headlights, and other devices which direct the light. The curved mirror behind the lamp can be directed to light up the area you are looking at. Those of us who are old enough may also recognise this shape from old electrical element heaters, where the back of the glowing element was silvered to direct the energy out towards the room.

To see videos on light and mirrors go to the companion website for the book at https://study.sagepub.com/chambersandsouter

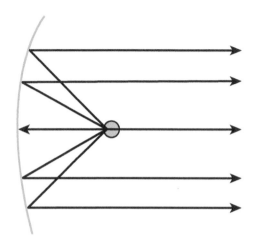

Figure 16.8

Photo 16.3

iStock.com/fullvalue

Light transmission in materials

Transparent, translucent or opaque?

It was shown earlier in the text that light travels in straight lines. Light can also travel through certain materials. Objects vary in how they transmit light. Transparent objects allow light to travel through them. Materials like air, water and clear glass are called transparent. When light encounters transparent materials, almost all of it passes directly through them. Glass for example is transparent to all visible light. Translucent objects allow some light to travel through them. Materials like frosted glass and some plastics are called translucent. When light strikes translucent materials, only some of the light passes through them. The light does not pass directly through the materials. It changes direction many times and is scattered as it passes through. Therefore we cannot see clearly through them and objects on the other side of a translucent object will appear fuzzy and unclear. Opaque objects block light from travelling through them. Most of the light is either reflected by the object or absorbed and converted to thermal energy. Materials such as wood, stone and metals are opaque to visible light.

Transparent materials

The most common transparent material is glass. It allows a great proportion of the light to pass through almost unaltered. Its use in windows allowed buildings to benefit from the light outside but not to lose heat. Original 'windows' were literally a hole in the wall. To reduce the heat loss the holes were covered with wood, cloth or animal hides. In the Far East, paper was used to cover large wooden frames which allowed a certain degree of light to enter the room.

There is evidence of some form of glass being used in Alexandria some two thousand years ago, but clear glass as we currently understand it for windows was not really available until the sixteenth or seventeenth century.

It was found that if we polish or cut the glass into various shapes it can affect the direction of the light from an image. These are generally known as lenses (from the Latin for lentil, as the lens was thought to resemble the shape of a lentil). Lenses have been recorded in early history but possibly as 'burning glasses'. (Lenses used to focus the rays of the sun to start a fire.)

Each of our eyes contains a small lens which helps focus the light we see onto our retina. Convex lenses can project an image of a distant object onto a screen.

As shown on the companion website, convex lenses can project an image onto a screen but the image would be different from what is seen. The image is upside down and the left-hand side and the right-hand side have been switched: we say the image is 'inverted'. The size and clarity of the image depend on a number of factors, such as the diameter of the lens and its curvature. Many devices employ lenses to help us in our lives: telescopes, binoculars, microscopes, cameras and data projectors all rely upon light being manipulated in order to help us see.

Concave lenses are more difficult to explain to young people but they are used in many devices.

Shadow formation

A shadow is formed when an opaque or translucent object is placed between a light source and a wall for example. A shadow is an area where the light reaching that area is less intense than the surrounding area. If we could block or stop all light reaching an area it would be totally dark or pitch black, but normally there is always light reflected off other objects to make the area not quite so dark. The most common shadows are those caused by the Sun. We can see our shadows in the morning when the Sun is rising. The shadows are long and get shorter as the Sun rises in the sky. Midday is generally when the Sun is at its highest point and this is when the shadows are at their shortest.

There are some good activities that can be done with a stick in the ground and marking its shadow. As seen on page 277 the Sun can draw an interesting path across the sky, and by marking its position at hourly intervals it can be easily seen. As said earlier, shadows are formed by an opaque object being placed between a light source and a screen. Simple activities can be used to illustrate this. Using their arms and fingers children can explore the shapes that shadows make on a wall or piece of card. The area a shadow takes up behind an object will be darker than the surrounding area.

Shadows obviously require a light source and sunlight from a window is the most obvious one. Torches and lamps however can provide a light source that will be different from the distant Sun. A torch can be used to project a light area upon a screen and this will have a different effect from that of sunlight. It will 'enlarge' the shadow whereas sunlight will not.

ACTIVITY

Making a simple sundial

A sundial is a device that uses the position of the Sun to give an estimate of the time of day. It does so using the relative position of the Sun on a scale of some sort. There is a wide range of sundials that can be explored relatively simply, but one simple method is to place a pencil or a piece of wood into the ground or possibly sticking out from an edge on a sun-facing wall:

- Place the stick or the pencil in a fixed position.
- Place a piece of card around the stick.
- Mark the position of the shadow at 10 am, midday and 2 pm.
- Indicate where you think 11 am and 1 pm will be.
- You have a rough sundial to give an indication of the time due to the position of the shadow being cast.

This is a simple but basic sundial. It will give a general indication of the time but is better suited for younger children.

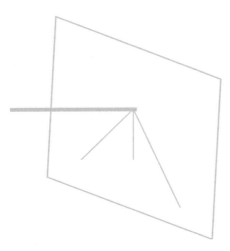

Figure 16.9

ACTIVITY

Making shadow shapes

Using card ask the class to draw a shape they like. Ensure the shape is no more than 10cm.

Cut out the shape and attach it with glue or tape to a straw.

Point the torch to a far wall and switch it on. Place the shape in front of the wall and notice the shadow it forms on the wall. Slowly move the shape towards the torch. Ask the class to observe the shape and size of the shadow as a pupil moves toward the torch.

Ask them to explain or describe what they observed.

They should link the shape they made with the shadow as being similar. They may also discuss the fact that the shadow is larger than the original shape. This could lead to an attempt to make shadow puppets or a small play based on the puppets. This is difficult but there is a long history of shadow puppets being used for storytelling and theatre.

Summary

This chapter provided an overview of a number of points around the nature and teaching of light and how we are able to see other objects. The topic lends itself to pupil experiment and investigation with ample opportunity for further study. There are some difficult concepts associated with this topic. For example, we can only see an object because light from a source has been reflected off the original object and then travels to our eyes. This concept – that it is the light that enters our eyes and not our eyes that do the 'looking' – can be difficult for young people. The structure of the chapter should allow you to improve your knowledge of the topic of light and the structure of our eyes, and also to combine these to highlight the nature of light energy for young people.

REFLECTION POINTS

- You ask your class to observe something in the playground and say to them 'Could you focus on the recycling bin?'. What is it they are actually focusing on?

  - They are looking at the bin in the playground. The light from the Sun is reflecting off the bin in all directions and some is entering our eyes. When we focus on it we are changing the shape of the lens in our eyes so that the light is being focused on our retina. We are not focusing on an external or distant object, we are focusing the light from that object internally. Corrective lenses help us focus the light on the retina.

- Why can we sometimes hear an emergency vehicle that is round a corner but not see it?

  - This may have more to do with how sound is propagated, but the key is that light travels in straight lines. If there is a building between us and the vehicle that is opaque we will not see it. Additionally, light from the vehicle may be reflected off a brick wall but brick is not a good reflector so we cannot see the image of the vehicle. Sound can disperse around corners and can also be reflected off walls, so we can hear the siren for example but not see the vehicle.

- A pupil stands in front of a wall and casts a shadow on a sunny day. Why is the area of the shadow not much darker as the pupil blocks the light from the Sun reaching the wall?

  - The shadow will indeed be duller than the rest of the wall but it will not be totally black. The reason for this is that there is light from all other surfaces being reflected. Some of this will reach the wall. The pupil only blocks the light coming from the Sun so the shaded section will be duller but not completely black.

17.
SOUND

Learning Objectives

Having read this chapter you should be able to:

- understand the basics of sound and its propagation
- understand how sound is generated by vibrating objects
- have become familiar with the concepts of pitch and volume and associated risks of loud sounds and noises
- understand how our ears detect sound

Overview

Sound is perhaps the main medium by which we communicate. Our conversations depend upon us being able to transfer sounds to each other. We can speak and address audiences of thousands using sound thanks to its ability to travel through the air. Young children spend the first few years of their life hearing a variety of noises and sounds which they learn to recognise and use to communicate with the world around them.

Sound's ability to travel across reasonable distances almost immediately and be heard by many people simultaneously allows us to teach, discuss and determine what others' opinions and viewpoints are. Teachers become effective public speakers using sound to explain and highlight key issues, and their skills of argument and instruction rely upon sound being spread across a classroom.

What causes sound?

In simple terms sound is caused by materials vibrating rapidly. On most occasions these vibrate so rapidly our eyes cannot detect the movement but our ears can hear the sound. A stick hitting the ground for example generates a sound, and we don't see the stick or the ground vibrating, but it does.

Molecules in the air are affected by these vibrations, and they vibrate and transfer the motions to neighbouring molecules, and it is this transfer of energy that distributes the sound through a room for example. As the energy from the source of the sound is spread throughout a classroom for example, it becomes less intense and as a result the volume decreases the further we are from the source of the sound.

What is the difference between sound and noise? This is more of a classification argument rather than a true, precise definition. In general people refer to sound or sounds as something they wish to hear. Noise is often referred to as unwanted sounds such as traffic in the background or something that stops us from listening to the thing we wish to hear. The term 'noise' stems from a Latin word meaning 'nausea' or something that makes you sick.

In simple explanatory terms for young children:

- Noise is a type of sound.
- Noise is sound we don't want to hear.
- Noise can be harmful to us.
- Sounds are more pleasant to hear.

As mentioned before, we often describe sounds we don't wish to hear as 'noise' and refer to the sounds made by machinery, aircraft, traffic, construction as 'noise pollution'. Noise is often associated with loudness, and in general noises are louder than sounds and this is where the harmful connotation is associated. Loudness is measured in Decibels (dB) (more of which later) and noise levels above 80dB can cause pain and long-term hearing difficulty.

Describing sound

Consider a simple note such as that created by a tuning fork or someone whistling. Strike the tuning fork off a bench then hold the base on a rigid object and it will produce a note that is simple or pure. The same is true for a simple whistle sound.

> To see videos on striking tuning forks of various pitches go to the companion website at https://study.sagepub.com/chambersandsouter

In describing how a high note differs from a low note we can use the terms pitch or frequency. The definition of frequency is the number of vibrations (or waves) per second and is measured in Hertz (Hz). Pitch is better for use with young children, but if you wish to compare notes in a slightly more accurate way getting the children to use the term frequency is worthwhile. High notes have a high frequency. A middle C has a frequency of 261Hz. The note A above middle C has a frequency of 440Hz and is used in tuning musical instruments. An octave is a range of (eight) notes where the frequency doubles. The note A above C has a frequency of 440Hz. The note an octave above has a frequency of 880Hz. The most common musical scales are written using eight notes, and the interval between the first and last notes is an octave.

Human hearing

Our ear is the organ we use for hearing. It has three sections: the outer ear, middle ear, and inner ear. The outer ear is the bit we see and is the curly section leading to the eardrum. The eardrum vibrates and transmits vibrations to the middle and inner ear which we detect as sound. We have two ears and this allows us to get an indication of where sounds come from as the volume and time delay between sounds reaching each ear can help us work out the direction. We can also move our outer ears slightly but we cannot turn them in the way that some animals do to detect danger. The outer ear is soft and easily damaged. If our ears are continually bumped or squeezed the cartilage becomes lumpy and misshapen. This is called 'cauliflower ear' and is common in boxers and rugby players.

Our ears can detect vibrations across a range of frequencies and this is generally accepted as ranging from 20 Hertz to 20,000 Hertz. In reality it is difficult to measure this and as people get older their upper limit of hearing drops dramatically. People over 30 will struggle to detect notes above 13,000Hz, but young school-aged children can easily detect these. As we age our hearing does degenerate but not for any single reason. In general it is the high frequency response that loses sensitivity first.

There are a number of apps for phones and tablets which are mini signal generators and oscilloscopes. They can describe the sound graphically, and are free and easy to use by all. They allow anyone to 'see' the image their voice makes.

Loudness

The loudness or volume of a sound is not a simple concept. We all know when a television or a song is played too loudly. If we go to a live concert the volume of the speakers can cause some damage to our ears and we will lose some sensitivity for a while. Occasionally this damage leads to a hissing in our ears which may take some hours to go away. Loudness is a subjective concept. It varies from person to person and is related mainly to frequency and sound pressure level. Additionally, if a sound is heard for a very short time it can appear less loud than the same sound played for longer.

In general terms the Decibel scale dB is used as a measure of how loud a sound is. Technically it is a measure of the sound pressure level and is useful in giving comparisons between sounds, although there are some technical difficulties associated with this.

Our ears have the ability to detect sounds which vary immensely in loudness. We can detect incredibly quiet sounds and also withstand very loud and powerful sounds. The power range from a quiet to a very painful sound varies by a factor of around 10,000,000,000,000. The scale is so large that to ascribe a number or value to it makes its use very difficult.

The Decibel scale is logarithmic and can give a value for loudness which can be appreciated, but it is complex. Again there are free apps for phones and tablets which will give you a reading for different sounds and noises.

An indication is given in Table 17.1.

The scale is logarithmic which means it increases by a *factor* of 10 when the dB level increases by 10.

A sound of 40dB is *10 times* louder than a sound of 30dB.

A sound of 60dB is *100 times* louder than a sound of 40dB (10 times 10).

Consider if you sang a note loudly and it was 70dB, that we then somehow brought in an exact copy of you, and that person stood beside you and sang at 70dB also. This would give a combined loudness of 73dB. It's a complex scale!

Table 17.2 gives an indication of the exposure times for Decibel levels.

Table 17.1

Weakest sound heard	0dB
Whisper – quiet library at 2m	30dB
Normal conversation at 1m	60–65dB
City traffic (inside car)	85dB
Pneumatic drill at 15m	95dB
Level at which sustained exposure may result in hearing loss	*90–95dB*
Hand drill	98dB
Motorcycle	100dB
Loud rock concert	115dB
Pain begins	*125dB*
Even short-term exposure can cause permanent damage – loudest recommended exposure WITH *hearing protection*	*140dB*
Jet engine at 30m	140dB
Death of hearing tissue	180dB
Loudest sound possible	194dB

Table 17.2

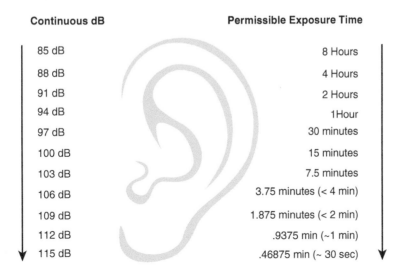

Continuous dB	Permissible Exposure Time
85 dB	8 Hours
88 dB	4 Hours
91 dB	2 Hours
94 dB	1 Hour
97 dB	30 minutes
100 dB	15 minutes
103 dB	7.5 minutes
106 dB	3.75 minutes (< 4 min)
109 dB	1.875 minutes (< 2 min)
112 dB	.9375 min (~1 min)
115 dB	.46875 min (~ 30 sec)

Hearing loss

There have been a number of studies recently in the United Kingdom and abroad which appear to indicate that Noise Induced Hearing Loss is increasing amongst young people. A study by Siemens in New Jersey concluded that as many as one in six young people are showing symptoms of hearing loss which is possibly permanent. This is generally accepted to be 30% higher than in the 1980s and 1990s. It is suggested that they have been undertaking activities which add to or exacerbate this loss.

The main 'activity' is listening to music with MP3/MP4 players, and in particular the use of in-ear or earbud-type headphones. Modern players can produce sounds up to 120dB for the person using them. The fact that the buds are inserted into the ear canal reduces the distance between the speaker and the eardrum, and this generates very loud noises in a small area inside the ear. Placing the earbud this close can increase the power by nearly 10 Decibels. There are two types of damage that can occur from this. Loud sounds damage the hair cells that respond to certain frequencies and the perception or ability to detect these frequencies is diminished. The high frequency (~6kHz) hair cells are the most sensitive, and they are the ones that are generally damaged first.

If the damage is so bad that the hair cells cannot repair or heal themselves, then the hearing at these frequencies is permanently diminished or the damaged cells fire continuously and you hear a constant noise. This is the cause of tinnitus. It can be loud and can seriously compromise sufferers' quality of life.

If you can hear the music playing from another person's music player then the odds are it is too loud and is causing some form of damage.

Many mobile phones and players have a settings control to limit the volume, but most people set this to maximum and do not set a lower volume.

The key to this is moderation. If the player is loud then the time exposed to that should be limited. If the volume is low then it can be played for a long time.

Table 17.3 gives an indication of safe hearing times for various volume settings.

A number of bodies give a 60:60 rule for using earbud headphones (i.e. it should be played at 60% of its volume for 60 minutes per day).

If you must listen to music it is safer and generally better quality if you buy a set of on-ear headphones, as the speaker is outside the ear and there is a degree of isolation provided by the over-ear padding, meaning you do not need it to be so loud in order to hear the music.

If you can afford it, noise-cancelling headphones are even better. They reduce the noise from any background source quite dramatically. The headphones detect the sound around the headphones and generate a similar but 'out of phase' noise which then cancel each other out. The headphones make the listener feel there is no background noise, and as a result there is less need to keep the volume high.

Table 17.3

% of volume control	Maximum listening time per day		
	Earbud	Isolator	Supra-aural
10–50%	No limit	No limit	No limit
60%	No limit	14 hours	No limit
70%	6 hours	3.4 hours	20 hours
80%	1.5 hours	50 minutes	4.9 hours
90%	22 minutes	12 minutes	1.2 hours
100%	5 minutes	3 minutes	18 minutes

Maximum listening time per day using NIOSH damage-risk, criteria.

Supra-aural headphones rest on top of the ear.

Source: US National Institute of Occupational Safety and Health (NIOSH)

Sound activities – making noise

An interesting 'toy' to make involves the use of a long spring and a card/plastic tube at one end. These are called 'thunder makers'.

- Take a cardboard tube with a plastic end (e.g a used coffee tin).
- Attach a length of string (about 50cm) to the plastic end of the tube.
- Hold the tube and vibrate it slightly. The vibrating string causes the plastic end to vibrate. This then causes the air in the drum to resonate, creating an incredible thunder sound which carries on for some time. Varying the size of the tube and the length of the string can provide a strong set of activities for investigating sound production.

Chimes

Chimes are simple to make and their sounds are generally pleasant. A simple and interesting take on this is to use a range of pencils that have already been used. Pencils with an eraser on the end are suitable as it is easier to insert a thread through the rubber section. Combining a range of pencils of different lengths suspended by a thread can produce a range of high and low sounds. Hanging these or little pieces of wood from a branch or an outside section of school will allow them to be moved by the wind, and hopefully provide some gentle sounds.

Making sounds and music

This is very practical area for children. Almost every material will make a sound if struck. Some materials such as glass or concrete are very fixed or rigid. When struck they will vibrate but the vibrations will stop very quickly. Those sounds tend to be quite sharp and sudden which is not really useful for music.

Materials which are able to produce a sound which can extend to a second or two are helpful here. Metals of the correct shape and length can vibrate and make a sound for a few seconds. A steel rod for example does not vibrate for long, but if we use the correct metal and form it can produce a sound for some time not unlike a bell, a tuning fork, or a chime.

These 'instruments' are generally described as percussive which means their being hit or struck. Children will note that some are (or could be) musical or tuneful, and classifying various instruments according to how they could be used can generate useful talk around the nature of sound or music.

Percussive instruments are believed to be the oldest group of instruments. Drums and sticks were used for rituals and dance in cultures possibly around 100,000 years ago. They were probably made to create beats and rhythms in ceremonies. Each of these relies on the principle of striking objects together, and this sound is combined with others in simplistic beats.

Other 'instruments' depend on air vibrating but do so in a range of ways.

Wind instruments

A large group of instruments, which can be tuned, work on a slightly different principle. Instead of vibrating a piece of metal or wood, wind instruments rely on a source of wind (our breath) to blow air into a tube or cavity. The air in the cavity vibrates and produces a sound which is then able to be heard.

By opening and closing holes with our fingers we can alter the frequency of the sound produced. In effect this produces a range of notes, allowing us to construct something melodic.

Our own 'blowing' provides the energy for the sound to be produced, and as long as we continue to blow or supply a flow of air that sound will continue to be produced.

There are many easy ways of making a simple whistle out of paper but a really straightforward one is by using a straw:

1   Take a straw and cut it so that it folds over but is not cut all the way through.
2   Fold it over so that it is at 90°.
3   Blow through one end and cover the other end with your finger.

The straw will make a whistle sound. Learners can investigate the sounds made by this by varying the length of the straw, where the cut is made, and by opening or closing the bottom end of the straw.

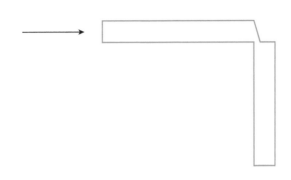

Figure 17.1

Stringed instruments

ACTIVITY

Making a stringed instrument

It is relatively simple to make a stringed instrument. A piece of wood, two pins, and some string or wire tied to the pins will make a basic instrument. The string can be hit or plucked and it will vibrate and produce a fairly basic sound. Thick wire or string will produce a low-pitched sound and thin wire will produce a higher-frequency sound. Additionally the tension of the wires will alter the frequency. A small wire or one that is tightly stretched will produce a high-pitched sound.

ACTIVITY

Building a stringed instrument

You will need a piece of wood about 30cm in length (although longer pieces will work well also).

Tie the string to one end of the piece of wood.

Place a pencil in the middle of the wood and pull the string down over the other end as shown in Figure 17.2.

(Continued)

(Continued)

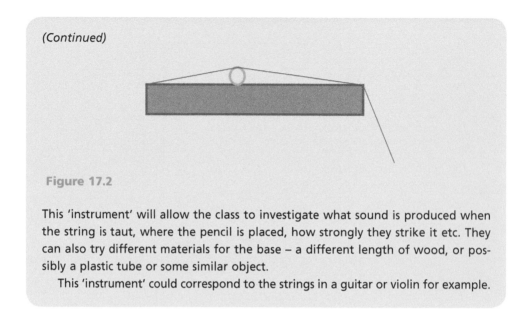

Figure 17.2

This 'instrument' will allow the class to investigate what sound is produced when the string is taut, where the pencil is placed, how strongly they strike it etc. They can also try different materials for the base – a different length of wood, or possibly a plastic tube or some similar object.

This 'instrument' could correspond to the strings in a guitar or violin for example.

Musical instruments such as guitars or violins also have large bodies of various shapes and sizes. These 'bodies' are vibrated by the strings and the way the strings are connected to them (e.g. via the bridge). These in turn vibrate and add to the sound produced. The precise acoustics of how they operate is very complex, but the shape, size, material and any holes all contribute to the overall unique sound of the instrument.

Noise pollution

Noise pollution is generally referred to as unwanted sounds that come as a function of traffic, passing aircraft, and similarly noisy activity. While at a distance the noise is not painful, the persistent background noise makes people unwell and irritable, and can affect their concentration and sleep patterns. There are many studies which link noise pollution to cardiovascular problems.

The CE Delft study on Traffic Noise Reduction in Europe, den Boer and Schroten (2007) concluded that 50,000 people die in Europe prematurely each year from heart attacks caused by traffic noise, with an additional 200,000 suffering from cardiovascular disease. The Transport and Environment website (www.transport environment.org/) is a great source for information on traffic and how we can deal with its impact.

Aircraft noise

Aircraft produce loud noise from a number of sources. The driving of the fuselage through the air pushes the molecules aside. This creates strong pressure waves and these spread out from the aircraft. This is affected by the speed of the aircraft and gets very loud at high speeds (and also at low levels as the air is denser). Low-level, high-speed passes by military aircraft for example can produce very loud noises.

A jet engine itself also produces very loud noise. The jet of air leaving the engine is travelling extremely quickly. This generates huge ripples in the air, leading to loud noise. This is responsible for the greater portion of the noise.

Traffic noise

Traffic noise is a combination of noises caused by vehicle engines, exhaust issues, tyre noise, and the vehicle pushing through the air. In general, the faster the vehicle travels the greater the noise. Many modern electric cars are virtually silent at low speeds and pedestrians often do not hear them when crossing the road. There have been suggestions to fit electric cars with a speaker system to generate noise so they can be heard by the general public.

Tyre noise is a main source of traffic noise and is caused by a number of factors. The tread rubbing against the ground, driving the car, is an obvious one. Different treads make different noises as the air is compressed into them as the tyre turns. Tyre firms have also attempted to design treads to cancel out the noise from other treads.

The air in the tyre gets compressed and acts like a large drum, reacting to different surfaces or holes to produce various hums or clicks.

Motorways are designed to be far away from residential areas in order to reduce noise pollution. Barriers are built at either side of the motorway to baffle the noise. Having trees and bushes along the side of the motorway also reduces the noise. Many motorways have roadside verges and slopes, and they also reduce the sound emanating from the traffic.

Sound and its propagation

As indicated earlier, sounds travel through the air due to the vibration of the particles. As the air particles vibrate they collide with other air particles next to them, and energy is transferred and spreads all around. This spread of energy results in the sound not being heard over long distances. The sound spreads out in all directions. It is similar to a stone splashing in a pond; it sends out a wave in all directions. A bell or a horn can create a loud noise but will fade quickly if it is in an open area. We can direct sound to some extent by trying to point the loudspeaker in a certain direction.

As sound travels through the air it strikes objects and some of the sound is reflected from them. Additionally, the sound can travel around barriers and corners a little, so you can hear sounds such as doors closing or ambulances travelling towards you but not locate these in your direct line of sight.

Concerts are generally in a hall or amphitheatre which reflects or guides the sound out towards the audience. Interestingly recent studies regarding Greek amphitheatres suggest that the limestone terraces absorbed certain lower frequencies which made the speech of the actors and musicians easier to hear. The main 'noise' or unwanted sound tends to be at the lower-frequency end and materials helped reduce this to leave a clearer sound. Modern theatres use amplification systems but the back of the stage generally reflects more of the sound out towards the audience, especially if it is a circular or parabolic shape.

This parabolic shape is used to pick up distant sounds and focus or point them to one area so they can be heard. Technicians listening for animal sounds often use a reflector attached to a microphone to detect distant animal sounds for example.

An example of such a device is shown below. These are relatively cheap and are freely available.

Photo 17.1

Photo by Leah Monson, U.S. Geological Survey, 2004

Devices similar to the one in the photo were also used to detect enemy aircraft during the First and Second World War. Radar had not been invented during World War I and was in its infancy during World War 2. Large parabolic reflectors were built on our beaches and soldiers stood in the centre, straining to hear the sound of the enemy bombers from over the horizon.

Photo 17.2

By Paul Glazzard, CC BY-SA 2.0, https://commons.wikimedia.org/w/index.php?curid=1667862

Hearing aids

People who are losing their hearing sensitivity can have hearing aids fitted. These are small devices which contain a microphone, amplifier, and loudspeaker. They fit tightly into the ear and detect outside sound, amplify it, and then play it through a small speaker inside the ear. Prior to electronic devices being available, people used large horn-shaped devices which gathered the sound over a large surface area and reflected it down to a small tube which was placed at the entrance to the ear. These were referred to as hearing trumpets. Some were very large and cumbersome.

Photo 17.3

By http://wellcomeimages.org/indexplus/image/M0013745.html, CC BY 4.0, https://commons.wikimedia.org/w/index.php?curid=36360515

Children can make a very basic hearing amplifier by taking a piece of paper and rolling it into a cone shape, with the narrow end having a hole that can cover the ear.

Roll the paper into a cone shape so that the small opening is about 5cm in diameter. Tape the edges of the paper together so the cone does not slide apart.

This cone is then pointed towards a distant sound so that people can hear it more clearly.

The small opening should be placed over the ear, not in the ear. Nothing should be placed in the ear.

All of these devices are similar to people cupping their hands around their ears. This reflects the sound from a larger area into a smaller point which makes it easier to hear.

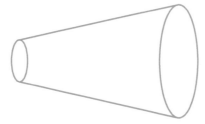

Figure 17.3

Summary

This chapter described in simple terms how sound is propagated and how our ears convert the air vibrations into signals we can understand. The separation of sound and noise was intended to allow us as teachers to highlight how what we want to listen to is often different to what we actually hear. It then led on to the associated health risks of using modern audio devices incorrectly. It is hoped that the basic descriptions of musical instruments, their operation and how we can build them, will allow an interdisciplinary approach where children can make music but also appreciate how the instruments create the sound and what we can do to describe this clearly.

REFLECTION POINTS

- Explain to a young person how they can hear their friends calling them from across the room.

  o When their friends call their name they use their mouth, throat and lungs. Their lungs push out a flow of air and this air vibrates the vocal chords and voice box. They use their mouth and throat to form certain sounds. The vibrations from their mouths are then transmitted by the air across the room. These vibrations reach the ears and this is converted into electrical impulses and sent to the brain.

- Why can in-ear headphones cause more hearing damage than on-ear headphones?

  - In-ear headphones have the speaker very close to the eardrum. As this distance is so small the sound is very loud. Small headphones can have a loudness of over 80dB. This can cause hearing damage. Users also tend to have the volume high so as to drown out the unwanted noise from their surroundings. As the sound level is concentrated within the ear it will not irritate other people and so they can listen for long periods. On-ear headphones block out unwanted noise so the music need not be played so loudly as the speakers are not so close to the eardrum.

- What is the difference between sound and noise?

  - Sound is what we want to hear. We use sound to communicate, play music, and interact with the world around us. Noise is sound that is caused by our everyday activities: doors slamming, machines turning, traffic passing. We generally do not want this noise as it does not convey information and will make it difficult to communicate.

Learning Objectives

Having read this chapter you should be able to:

- have a good understanding of what forces are and what they can do
- explain what magnets are and how we can use them
- describe what friction is and how it acts on moving objects
- understand how simple machines operate

Overview

In order to explain the physical world around us, and in particular how things move and are shaped, we use the concept of a force. We can explain the motion of things by examining the forces acting on an object, and we can describe what happens to the shape of objects when a force acts on them. You will notice that we refer to a force 'acting on'. This is slightly technical language and it can refer to a situation when you squeeze a sponge ball for example or drop an object. It covers situations when an object is in contact with another object, and others where there is no contact (magnets for example). A force can still be acting in either case.

The difficulty and confusion around forces and motion are not new. The Greek philosopher Aristotle (384–322 BC) is generally credited with making the first real attempt to understand or explain the motion of objects when moving. He wrote a series of books and papers on a wide range of topics in a great attempt to explain the natural phenomena of the world. Much of his writing had a philosophical position, and, he believed things had their 'natural place' and that there were celestial and terrestrial elements which could combine and behave according to certain rules. To fully understand the principles and reasons or explanations of Aristotelian science is difficult, but it was (and is) taught in many universities as part of the way of training students in rational or analytical thinking.

In explaining falling and rising objects it was proposed that all objects move toward their natural place. Earth and water are on the ground or fall to the ground because that is their natural place. Bubbles rise in water and flames move upwards because air and fire have a 'higher' natural place. On explaining the motion of objects Aristotle observed horses and carts moving across roads. Once a horse is removed from a cart it stops. Aristotle deduced that in order for objects to move a force must be applied continually. This led to the belief that objects slowing or having to overcome some form of resistance was the natural way of things. There were one or two examples where the explanation needed more detail. The classic example was that of an arrow being fired from a bow. The arrow would be fired and travel huge distances before coming to rest. What enabled it to travel so far when some sort of resistance should have slowed and stopped it? It was proposed that the arrow was moving forward so quickly there was a small vacuum at the back of it which drew the air in. This then struck the arrow pushing it forwards. Again, wrong but plausible.

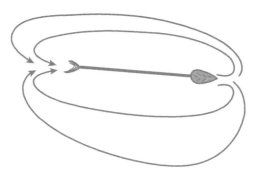

Figure 18.1

These examples are not provided to increase your knowledge of ancient Greek philosophers but to give an indication of the difficulties associated with this topic. Young children will observe things moving and falling and construct their own ideas of why those events occur. The teacher's role is to guide them towards a clearer understanding of why these events occur. It was the work of scientists such as Galileo, Copernicus and Newton to really explain what a force can do and how it affects motion.

In simple terms a force can change the speed, shape or direction of an object. We can tell that a force has acted by observing a change in speed, shape or direction.

Changing speed

A simple example of this would be dropping a ball. As the ball is released it falls to the ground and as it falls it speeds up. It is changing its speed; therefore a force is acting on the ball. The same happens if we press down on the accelerator in a car. The force from the engine makes the car change it speed.

In all of the examples above the applied force made the object change its speed quite dramatically. The same sort of change in speed will also occur when we are driving or cycling and apply the brakes. The brakes apply a force to the wheels, and in turn to the vehicle, and this changes the speed of the vehicle (i.e. it slows down). We interpret this as a change in speed, and therefore a force is acting. Another obvious example of this is a parachutist opening a parachute. This creates a very large drag force and slows the person down dramatically. People mistake this slowing down as the parachutist appearing to move upwards. This is incorrect. The camera operator is still falling at a high speed and the person who has opened the parachute is slowing down. The camera operator is still travelling at a high speed and in the camera's view the person is moving upwards.

The same is true of drag cars and the retired Space Shuttle. Once the race was over or the shuttle made contact with the runway a parachute was deployed and this slowed the vehicles down until they eventually stopped.

In lower primary, activities that could be used to reinforce or consolidate children's ideas in this area would include asking them to describe situations where they have observed objects being moved quickly or slowed down quickly. Worksheets or images of aeroplanes taking off, racing cars, playground chutes and roundabouts, buses, large trucks, falling objects etc. can be shown.

These situations can be used to examine the reasons why objects move quickly or slowly. It is acceptable to use the terms pushing and pulling in this context. Many teaching guides promote this, and as an introduction to the idea that in order to move things (speed up or slow down) we apply a push or a pull.

These activities describe the nature of a force in terms of it being in contact with the other object. We would suggest that you introduce the words 'in contact' in these situations. This also allows children to talk about large and small forces. For example, the force needed to open a large door is less than the force required to open a desk or a cupboard door. There are small, inexpensive pieces of apparatus known as Newton balances which are simple spring balances that measure the force required to do something but measure it in Newtons.

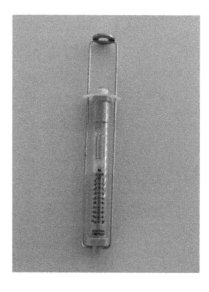

Photo 18.1

ACTIVITY

Measuring forces

Using the Newton balances ask the class to measure the force required to:

- lift their school bag;
- open a cupboard door;
- drag a bag across the floor;
- lift a book.

This introduces the unit of force, the Newton. This was named after Sir Isaac Newton and it was he who was credited with changing the old Aristotlean view of the world. There were other scientists around at that time and before, who had suggested or proposed alternative views of how the world moved, but Sir Isaac Newton brought them together clearly. Newton alluded to the work done by others in a letter to another scientist, Robert Hooke, by writing 'If I have seen further it is by standing on the shoulders of giants'. This inscription can be found on the UK's £2 coin.

The spring balance allows children to measure forces in Newtons and they can begin to appreciate the size or magnitude of what the unit conveys. It is a small unit and the weight of a 100g object is 1 Newton (1N). If you have a set of scales you can determine the weight of the children by measuring their mass in kg then multiplying by 10 to obtain their weight. A child of 35kg will have a weight of 350N. (See Chapter 19 for additional information on weight and mass.)

Changing shape

A force can deform, twist, bend or stretch an object. This can be exemplified by squeezing sections of sponge, stretching springs, or dropping objects onto soft wood or plasticene. The children can see the object deform and in many cases also see the extent to which it has changed its shape. Learners should be exposed to examples of objects changing shape: eggs being broken, paper being folded or cut, a balloon being squeezed and released etc. In examples such as these they can be asked their ideas on what the force is doing, or if the change in shape was large, or whether the object returned to its original shape.

A force will cause any object to deform or change shape. *Any object*. As you walk across a floor or a pavement, the floor is depressed downwards but by only an incredibly small amount. When you place a coin on a table, the table will 'sag' in the area around the coin. This is the fuller meaning of 'A force can change the shape of an object'. It changes the shape of any object, not just soft ones. Consider walking across a soft mattress. It deforms and curves down where you stand on it. Change it to a firm mattress and it curves in a general shape but not as much. Walk across a wooden floor and it will deform but you will not notice it. This is what occurs every time an object moves across another one. They change their shape. Buses driving along a road gradually deform the road.

This is very difficult for children to appreciate and what was explained in the paragraph above is far beyond what you should teach. It does however allow you to consider children's misconceptions in this area and where these can arise.

Changing direction

A force acting on an object can cause it to change direction. Roll table-tennis balls along a table and allow the balls to collide with each other. It can be seen that the balls bounce off each other. They then move across the table, backwards or sideways depending upon the nature of the collisions.

Roll the table-tennis balls across the table and allow the children to blow them off course using a straw or a rolled-up piece of paper. Although not visible, the children can be encouraged to describe the change in direction being caused by the air 'pushing' these off course.

Discuss a clip of a football or rugby match. Describe what happens to the ball and the players as they collide with each other. Some will go sideways, some backwards. A tennis match is a perfect example of balls changing direction continuously due to the many forces being applied to them. Cars change direction by turning their wheels and this creates a sideways-acting force which turns the car.

All of the above examples have been described by the action of forces which are in contact with an object. The children can physically see the other object in contact with the initial object and observe the consequences of this.

Forces at a distance

A force can act on an object even when it is not in direct contact. The most obvious example of this happens when an object is falling. As it falls it gets quicker, it changes its speed so we can say a force has acted. Gravity!

If a ball is thrown upwards it will slow down and stop for a brief point at the top of its flight, then change direction and come back down to Earth, increasing its speed as it falls.

Another simple example of this is rolling a ball off a bench. Once it leaves the bench it does not travel horizontally, it starts to move vertically and gradually goes from being horizontal to vertical. It is changing its direction. Therefore a force has acted on it.

Magnets and electrically-charged materials can exert forces on objects without being in contact. Magnets can cause ball bearings to turn and change direction. A compass will always turn 'northwards' if it is rotated away from North.

Forces 'acting at a distance' are more difficult for young children in that they see no concrete mechanism causing the object to float, move, or be attracted. In these circumstances describing what they see may need greater structure. Assisting learners with headings such as 'What moved?' or 'What direction did it go in?' will point them towards the desired outcome.

Magnets

The topic of magnets is naturally of interest to most people, especially young and inquisitive children. The ability of a seemingly normal object or piece of metal to be able to attract other 'non-magnetic' things, to stick to fridges, to push other magnets across a table or to make objects float in mid-air is intriguing. The questions these phenomena pose include:

- What is a magnet?
- What makes something magnetic?
- How do we make magnets?
- Where do we use them?

Magnetism, in its common sense, is due to the structure of the atom. In most cases the magnetic properties of atoms are randomly spread so that they in effect cancel any overall magnetic effect. In certain materials the magnetic properties are 'lined up' differently, and it is this arrangement that leads to the magnetic materials we are familiar with. The mechanism which leads to the common, everyday permanent magnets that we are familiar with is complex. It is enough to say that some materials are 'magnetic' (i.e. they have an area around them which can affect other objects if brought near). We can show this by bringing an object close to the magnets and observing what happens.

To see video clips of magnets in action see the website at
https://study.sagepub.com/chambersandsouter

People have known about 'magnetism' for thousands of years. Ancient Greeks and Romans were aware of the properties of materials known as 'lodestones' which could attract other pieces of iron. The ancient Chinese built elaborate items using lodestones with compasses in them in order to align their houses correctly (Feng Shui).

Activities with magnets

The most obvious activity is investigating which materials can be attracted to a permanent magnet.

The main metals that are attracted to magnets are iron, nickel and cobalt. Many other metals may contain some of these materials. As a result they may be attracted to a magnet. It is not important that they know and can recall which materials are attracted, just that some are. Glass, wood, wool, paper etc. should all be tested to provide a more comprehensive list of materials.

A key part here is that magnets are attracting (applying a force) to other objects. Whilst the object may be attracted to the magnet, the force attracts the object to the magnet. The magnet does not actually need to touch the object for it to stick. The object is attracted and pulled towards the magnet. The force is acting at a distance, albeit a small one.

Making magnets

Iron is a strong material in magnetic terms and many everyday objects will contain iron to some extent. You can magnetise small pieces of materials containing iron by placing a magnet in contact with the material. An iron nail can become magnetic in this way. If the piece of iron is kept in contact or rubbed against a magnet for a period of time, the iron can remain magnetic even when the magnet has been removed. It tends to remain as a weak magnet however and gradually fades.

Industrial magnets are made in a complex process. The main component metals such as cobalt and nickel are heated and melted to form alloys. These are then ground into a fine powder and compressed or moulded within a magnetic field. They are then machined into the correct shape, heated, and passed along a very strong magnet.

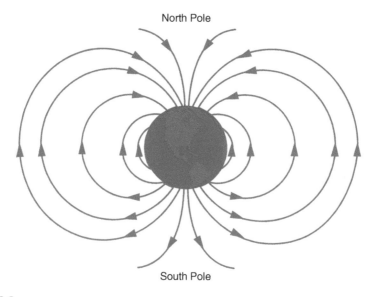

Figure 18.2

Magnets are used in an incredibly wide variety of components:

- Electrical motors, including microwaves, washing machines, electric cars, kitchen accessories, lawn mowers and the like.
- Loudspeakers and headphone sets.
- Mouthpieces for microphones.
- Computers.
- Magnetic latches for cabinet doors.

Making a compass

The Earth is magnetic. It has a large magnetic field which surrounds it. It is accepted that the cause of this magnetic field is due to the molten iron and nickel in the Earth's core. It is a relatively weak magnet. Other smaller magnets will align themselves with the magnetic field and given the chance will rotate and point to magnetic North. We can make a compass by placing a magnet on a frictionless surface and allowing it to rotate towards North.

Friction

Friction is a force that acts when two materials come into contact with each other. Young children may be familiar with the word 'friction' and use it in an everyday sense. When children are asked to rub their hands together they can describe this as 'friction makes my hands warm' or such. This is a good description but doesn't really clarify what friction is. In primary-school terms friction can be exemplified by sliding blocks of various materials along a piece of wood or plastic. Placing a piece of wood on some books to make a slope and sliding blocks down allows children to investigate the nature of frictional forces. The movement of the blocks down the slope is related to a number of factors. The most obvious ones are the weight, smoothness and surface area of the blocks, and the smoothness of the surface of the slope.

How do we know friction is a force?

As said earlier, a force can change the speed, shape and direction of an object. Push an object along the floor and let it go. It will stop after a while. Friction has slowed down the object, therefore its speed has changed and a force has acted. The action of friction also changes the shape of an object. Look at the soles of your shoes after wearing them for few weeks: the surface will have become scraped, tread will have been lost, or part of the sole or heel will have worn away. The shape will have changed. Here the emphasis in teaching is to allow learners to identify situations where friction occurs and what its effects are. In general friction forces are greater when there are heavy objects, a larger surface area in contact, and materials have rougher surfaces.

Air friction and drag

Although not obvious in many circumstances, when an object passes through air there are frictional effects. Air friction has two main components. One is the friction due to the air rubbing against the surface of the object such as a car. The other is due to the way air moves around the shape of the car. If an object has a large square surface area, such as a bus or a square sheet of paper, the drag can have a large affect on the movement of that object.

An important point here is that it is air friction or drag which limits the speed we can travel in cars, buses, trains and aeroplanes. At low speeds drag is small but it increases very quickly as the speed increases. An indication of this would be if you were to place your hand out of the window of your car. At low speeds this is hardly noticeable, but at 30mph and above it is very noticeable. The force on your hand would bend your hand back quite severely. Consider now the force acting on the total surface of your car as it is moving. The frictional forces are great. It is these forces that limit the maximum speed of your car. If you have a small engine the force produced by that engine will be matched by the drag when it reaches a reasonably high speed. At this point the drag stops your car from going faster and it is this that limits the car's maximum speed.

You will notice this if you are driving on a motorway at around 60mph. Take your foot off the accelerator and the car will immediately start to slow down. This is the drag changing your speed, and therefore it is a force.

The same principle applies when aeroplanes fly through the air. A passenger jet travelling at 500mph can have all its engines running at close to maximum and it will not speed up. The frictional forces at this speed are great and counter or balance the force from the engines. Again, if the engines were to reduce their forces the aeroplane would slow down dramatically.

From this we can see that if we wish to travel quickly through air we can either use engines or machines that provide a large driving force *or* reduce the air friction. The name given to the principle of reducing air friction is 'streamlining'.

The general principle here is to make the shape such that it cuts through the air (or water) and to have the surface as smooth as possible.

Fast cars tend to have smooth lines and polished surfaces to allow this. However most modern cars also operate by these principles. An interesting point here is to compare the shape of a car from the 1970s with its modern equivalent and notice how the shape has changed.

Photo 18.2

By Allen Watkin from London, UK – Cortina Mk3, CC BY-SA 2.0, https://commons.wikimedia.org/w/index.php?curid=32033910

Photo 18.3

By Thesupermat – own work, CC BY-SA 3.0, https://commons.wikimedia.org/w/index.php?curid=23044241

ACTIVITY

Air friction

Take a small solid object such as a tennis ball and a sheet of paper:

- Drop both from a height of about a metre.
- The ball will fall down more quickly than the paper. Ask the children why this is and you should receive a range of answers, including its weight, shape, size etc.

(Continued)

(Continued)

- Crush the paper into a ball and ask the class to predict what will happen. Repeat the experiment. (Both will hit the ground at the same time.)
- Ask the children to explain the difference.

This simple experiment highlights the effects of air friction at small speeds. It provides an appropriate introduction to how can we reduce the effects of air resistance. What should we do if we wanted to travel quickly through air (or water)? Comparisons of fast- and slow-moving objects are very useful for highlighting the general principles of reducing drag. Sea-living animals such as fish and seals are particularly good examples, as they have to travel through water in order to survive and hunt. They have developed a natural streamlined shape.

Photo 18.4

By Albert kok – own work, CC BY-SA 3.0, https://commons.wikimedia.org/w/index.php?curid=6758553

Examining the shape of a shark for example will allow children to see the smooth curved shape that cuts through the water. This shape reduces drag as does the shark's skin.

Photo 18.5

By Pascal Deynat/Odontobase – own work, CC BY-SA 3.0, https://commons.wikimedia.org/w/index.php?curid=15564795

When examined closely a shark's skin has tiny 'scales'. These scales actually reduce drag in water and combined with the way a shark bends its body to produce thrust allows it to move very efficiently. Swimwear companies make swimsuits with a 'shark-skin' effect although some of these suits are banned from competition as they provide too much of an advantage for the wearer.

In general any sort of smooth, rounded, 'teardrop' shape will reduce drag. Cyclists' helmets, submarines, attachments to long distance lorries etc. are all designed around a smooth curved section as much as possible.

Photo 18.6

iStock.com/Christian LÃ¼ck

The image above shows the 'streamlining' of a large truck. We need the truck to be large enough to carry the load, but it has been adapted by adding a smooth curved section above the cabin to reduce drag, and therefore improve fuel consumption.

Difficulties with identifying forces

Many people struggle with conceptualising forces. As we said earlier, these are not seen and our everyday assumptions of how things act often do not hold up to scrutiny.

Place a book on a table. What forces are acting on the book? (An obvious one is the weight of the book.)

In what direction does its weight act? (Downwards.)

What other force is acting? (There is a force from the table on the book. This acts in an upwards direction.)

How do we know this second force is acting? Consider if the upwards force was not there. If the only force acting on the book was its weight then it would fall downwards. It isn't though so there must be another force in the opposite direction. This is the force from the table on the book. It acts up and balances the weight. Therefore the book does not move. Two forces have acted 'against' each other and have cancelled out each other's effect.

Drop a tennis ball

As the ball is falling what forces are acting on it? Its weight is the main one. This force changes its speed as it falls to the ground. There is some drag but it is much smaller than its weight so it accelerates downwards.

A helicopter hovering in mid-air

What forces are acting on it? Its weight is acting on it in a downwards direction. The force from the blades are providing a force in the upwards direction. The forces are of the same size so the helicopter will hover.

A boat on the water

The weight of the boat is acting downwards. There is also an upwards force from the water on the boat. When these are balanced the boat will float in the water. If more cargo is placed on the boat it becomes heavier and will move downwards until it displaces more water which increases the upwards force.

Examples such as these should help consolidate your own and possibly your class's ideas regarding forces. The identification of forces is difficult and possibly beyond the scope of primary schools, but as a teacher you need to be able to discuss the possibilities going beyond what they are learning. This confidence in your increased knowledge will allow you to explore situations that may arise when teaching.

Simple machines

A machine is a device that we use to help us do things more easily. In forces terms we use machines to lift or move things that are heavy or require a large force. Simple machines involve the principle of a lever.

A lever is a piece of wood or metal that exerts a force on an object. The force exerted may be greater than that which we can exert ourselves. A simple example of this is a hammer pulling out a nail. We couldn't pull the nail out directly but by using a lever we can.

A simple lever contains a fulcrum or pivot and something for us to apply a force to. The fulcrum allows us to apply a larger force to an object more easily

In the example above we press down and the lever lifts a heavy object. We do not get this extra force out of nothing however. We apply a smaller force but we push down for a longer distance. Our hands may push down a metre but the rock will only rise by 20cm for example.

It is easy to build a simple lever. Get a plank and piece of wood such as a brush handle to act as the fulcrum. Place the fulcrum closer to one end (3:1 is good for this). Place a heavy weight on one end and allow the class to see how difficult (or not) it

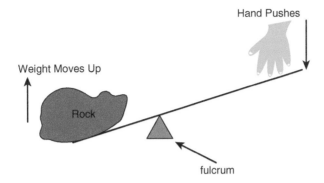

Figure 18.3

is to raise the weight. It is quite easy to raise a pupil standing on the far end of the plank using this method. We use a wheelbarrow to lift and move heavy objects. In this case the fulcrum is the wheel. The handles are a small distance from the wheel and we apply a force to the handles and lift the wheelbarrow. This raises the load in the barrow but only by a small amount, so we can lift heavier objects with it.

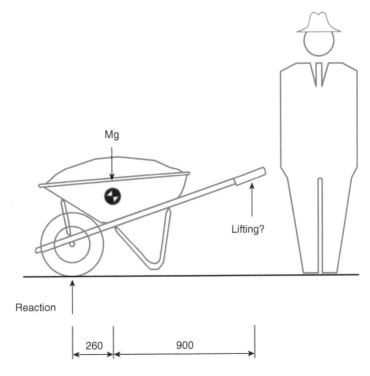

Figure 18.4

Using a piece of metal to open a tin of paint is an example of a simple machine. A screwdriver turns a screw more easily because the handle is thicker and this acts like a lever.

Summary

This chapter has highlighted a number of ways where we can explain everyday events using the concept of a force. It also clarified and explained in simple terms what forces can do and how we know they have acted. Real-life examples are often far more complex and to analyse these accurately can be counter-intuitive. The key here is to introduce the basic ideas in a simple and straightforward manner. Most incidents can be explained simply in general terms with forces changing the shape, speed or direction of an object. An important point is to accept that we cannot see the force itself, only what it has done or caused to happen. We can see a hammer hitting a nail but we cannot see the force itself.

REFLECTION POINTS

- How would you explain to a pupil that gravity is a force when it cannot be seen and does not touch an object?

  - You could respond by saying a force can change the shape, speed or direction of an object. If a ball is dropped it falls and gets faster as it falls. It is changing its speed. You could also throw the ball upwards and watch it come back down. It was going up but is now moving downwards. You cannot 'see' a force, but the ball has changed direction and speed and therefore a force has acted on it.

- Why do car tyres have to be replaced occasionally?

  - Tyres are required by law to have a certain amount of tread. They transmit the force from the engine and brakes to the road when the car speeds up and slows down. They do this by friction between the tyre and road. Friction is a force and can change the shape of an object. Each time the car moves the friction wears away a small amount of the tyre. Over time more is worn away until there is very little tread left. This is when the tyre needs to be replaced.

- What can we do in order to ensure things move more easily?

  - We can make them more streamlined so that we reduce the friction between the object and the air. We can do this by making the object smooth with no hard corners. This allows the air to flow over the object smoothly. Cyclists wear smooth helmets and lean over when cycling to allow the air to flow over their head and back smoothly. They also tuck their arms in to be more streamlined. The same is true for downhill skiers also.

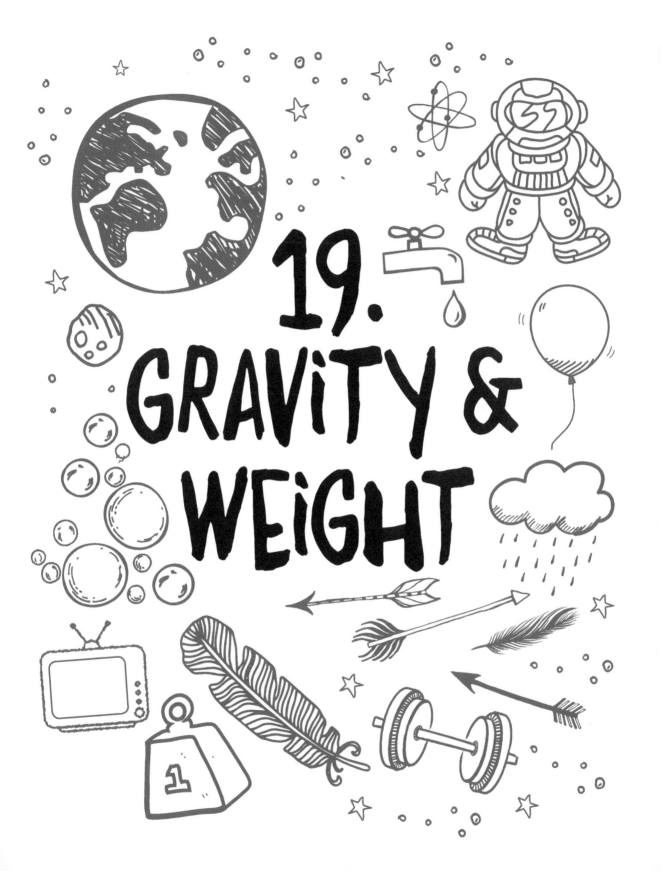

19. GRAVITY & WEIGHT

Learning Objectives

Having read this chapter you should be able to:

- describe gravity and understand how it can be related to children's experiences
- understand differences between weight and mass and how to demonstrate this using falling objects and light and heavy materials
- be aware of some common misconceptions regarding gravity and space, and explanations of the solar system, stars and planets

Overview

Gravity is the term we use to describe the attraction that masses have for each other. Any object which has mass will attract and be attracted to other masses by this process. Larger objects have greater mass and are therefore able to attract other objects to it more strongly. The Earth is a large, massive object that attracts us, the Moon, and any passing piece of space debris.

We are 'held on' the surface by the Earth's gravitational attraction. If we lift a stone and release it, it will fall to the ground. This is due to this attractive force between the stone and the Earth.

All objects are affected by the gravitational attraction from other objects. However our masses are so insignificant that the force of attraction between us and other small masses is so small it cannot be felt or measured.

Figure 19.1

The term 'force of gravity' is used much in general discussion regarding weight, objects falling and so on, but it is an imprecise term. It implies there is a single force, but this is misleading. The force experienced by a falling ball is different from the force experienced by an aeroplane flying at 10,000m.

Sir Isaac Newton is credited with the 'discovery' of gravity, but other scientists had also been working in this area trying to explain the motion of the planets, moons, and other astronomical objects. There had to be something that kept the planets and moons in orbit but it was difficult to explain what this was. Robert Hooke (1635–1703),

Galileo (1564–1642) and Kepler (1571–1630) were all concerned with the motion of objects which was apparently caused by an unseen force.

The exact mechanism by which we are attracted is still unclear and extremely complex! Various scientists at that time proposed waves or 'things' emanating from one object to another, but there have always been difficulties in attempting to explain gravity more fully.

For our purposes here we shall initially consider gravity as it appears on Earth in our everyday experiences and how it applies to learners' perceptions.

Gravity makes things fall

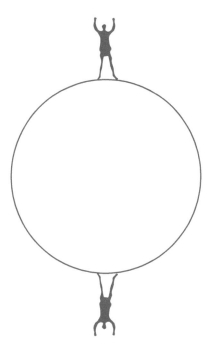

Figure 19.2

This is our starting point. We are attracted to the surface of the Earth by gravity. This is a simple and straightforward statement which all children will have concrete experience of.

We are 'pulled down' by a force which keeps us on the surface.

Simple experiments can be undertaken here where children drop stones, balls, coins and other objects into sand and they are asked to observe what they have seen. This can be expanded with older children to determine how they fall.

Do objects fall at a steady speed? Do objects speed up when falling?

Gravity is a force and when an object is acted on by a force it speeds up (i.e. accelerates). Dropping objects in class may not make this clear. You could use appropriate questioning with pupils to direct their own experiences to answer this:

- Would a toy be damaged if it fell off the desk or from a high shelf?
- What would be more uncomfortable – you dropping your schoolbag on your toe, or it dropping on your toe from the top of a building?

Simple classroom experiments can be done here, dropping marbles into soft sand or water. Dropping these from different heights would show that objects dropped from greater heights hit the ground at higher speeds.

> To view a clip of a classroom experiment on gravity access the companion website for this book at **https://study.sagepub.com/chambersandsouter**

These experiments may show in simple terms that falling objects increase their speed, but this will be in conflict with much of what they observe in everyday life:

- Leaves falling off trees do so slowly, they don't speed up.
- Balloons can appear to go upwards from being kicked or struck, and when they fall down they do so at a slow, steady speed.
- Dropped pieces of paper and lightweight objects fall slowly and at a steady speed in a classroom.
- Snowflakes fall slowly.

Each of these involves introducing another factor – air friction – which leads to situations where more than one force is acting, and thus makes the explanation more difficult. Helping children to give a correct answer when they cannot comprehend the full explanation is an issue for teachers. Accept their explanations as appropriate to their stage of development, but also be aware that these explanations will develop over time as the pupils do.

Do heavy objects fall more quickly than light objects?

All objects will be pulled to Earth by its gravitational attraction. The size of that attraction is determined by the size, or more correctly, the mass of the object. A large, massive object is pulled to the surface more than a lightweight object. The 'force of gravity' or the force exerted on an object by the Earth is its *weight*.

This seems somehow incorrect to many. Larger objects 'feel' more difficult to lift, so it is easy to assume that there must be a greater force pulling these down, and therefore they will fall more quickly.

ACTIVITY

Dropping plasticine balls

Make four small balls out of plasticene of roughly the same size.

Hold these out in front of you and release them at the same time. Note which hits the ground first. Repeat this experiment but drop the balls from different heights. They should all hit the ground at the same time.

Take three of the balls and squeeze these together to make a larger ball.

Repeat the experiment using the 'large' ball and the small ball.

Learners should note that these do not fall at different rates. A larger object does not fall more quickly than a smaller object, and this may help in justifying that heavy things do not fall more quickly than lighter things. It may not counter the children's experiences and misconceptions in heavy things falling more quickly, but it is an example of an experiment or phenomena which they have seen and have to process themselves.

Falling through the air

When an object is raised and released it falls to the ground. Gravity exerts a force on the object and so it falls and its speed increases.

Figure 19.3

As the object falls, it increases its speed. As it does so it has to move air out of its path. This process of air being forced around the object acts as a resistive or frictional force which makes it more difficult for the object to travel through the air. We call this 'resisting' force air resistance, air friction, or drag.

This air friction is a force which acts in the opposite direction. It can slow down an already fast-moving object and it will stop objects from falling at faster and faster speeds. Air friction stops it from falling more quickly. This acts against the downward motion of the object, against its weight.

A feather or sheet of paper is a lightweight object with a fairly large surface area. It has to move through a wide area of air as it falls. Moving through this air creates a large air resistance which in turn does not allow the paper to reach a high speed. This causes it to fall quite slowly.

This is the basis of the misconception that light things fall more slowly than heavy things. In general they do but the slow speeds are due to air resistance rather than their weight. If we could drop two objects in a vacuum, say a feather and a hammer, these would hit the ground at the same time.

A tennis ball for example can travel at quite a speed before the air friction prevents it from going any faster.

Skydivers can exit an aeroplane and reach a speed of about 54 metres per second (122mph). They do not fall faster than this speed. It is not that reassuring to know that you would not go faster than this if you were to fall from a great height.

In this situation the force of gravity pulling you to Earth, your weight, is balanced by the air friction acting against your fall. This does not slow you down, it stops you going more quickly. This is referred to as 'terminal velocity'.

Figure 19.4

By Julio Gomez Piqueras (own work) [CC BY-SA 4.0 (https://commons.wikimedia.org/w/index.php?curid=44718986)]

The terminal velocity for a balloon is very slow because the air friction needed to match or counteract the weight is very small. This is why balloons, feathers and pieces of paper fall slowly. It is not solely because they are lightweight.

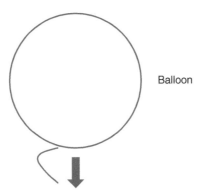

Balloon

Figure 19.5

It is fortunate that all falling things have a terminal velocity. Raindrops have a terminal velocity of about 4 ms^{-1} which is good for us. Raindrops fall from great heights and if they did *not* reach terminal velocity they would be hitting the ground at speeds which could cause us injury.

When a skydiver opens their parachute their falling or terminal velocity goes from about 54ms^{-1} to about 5ms^{-1}. This allows them to land safely. When their parachute opens the air friction increases dramatically due to its large surface area.

In this case skydivers *do* slow down. This is the cause of another common misconception. When a parachute opens the sky diver does not go quickly upwards!

Generally when filming these events there are two skydivers. Both are falling at around 54ms^{-1} and they will have cameras attached to their helmets.

Skydiver 1 is filming Skydiver 2. Skydiver 2 opens the parachute and slows down very quickly. Skydiver 1 is still falling at 54ms^{-1}, but to the view on Skydiver 1's camera Skydiver 2 has appeared to have zoomed upwards. This gives the impression that opening a parachute makes a skydiver suddenly move upwards.

Weight and mass

In scientific terms the weight of an object is the attractive force a large object (a planet or moon for example) exerts on a small body. Our weight is the force attracting us to the planet's surface. Like all forces it is measured in Newtons (N).

It is important to explain the difference between weight and mass, however this is muddied by a lack of familiarity with the unit, a belief that our weight is measured in 'stones' or 'kilograms', and the difficulty of discerning the difference in a classroom situation when weight and mass are virtually identical.

What is mass?

It is simply the amount of matter that an object is composed of. For us that is skin, bones, organs, fat, blood and so on. The mass of an object is measured in kilograms (kg) or grams (g). The mass of an object does not always depend upon its size. A block of polystyrene will have a much smaller mass than a block of steel for example.

Is mass related to weight?

Yes. The weight of an object is the gravitational attraction the Earth has on that mass. On the surface of the Earth these are very closely linked. The weight of an object can be calculated by multiplying its mass by the strength of the gravitational field at that point. On the surface of the Earth the gravitational field strength (g) is taken as 10 (Newtons/kilogram).

The weight of a 24kg object can be calculated by multiplying its mass by 10.

$$\text{Weight} = \text{mass} \times g = 24 \times 10 = 240\text{N}$$

The scientific or correct unit of weight is the Newton (N), named after Sir Isaac Newton. When engineers calculate the loads on buildings and other structures they do so in Newtons or kilonewtons (kN), not tons or pounds. Before the Newton was accepted as the unit of force engineers used strange units such as pound mass and pound force (and even a unit called a 'slug') to distinguish between these. The SI system makes this distinction more straightforward. Mass is measured in kilograms. Weight is measured in Newtons.

How to lose weight

Your mass is a measure of the amount of matter or material you are composed of. Your weight is the force that the Earth attracts you with or holds you to the surface with.

Your weight in Newtons can be calculated by multiplying your mass by ten. If you have a mass of 55kg your weight is 550N. Interestingly the gravitational field strength on the Moon is roughly one sixth of that on Earth. Your weight on the Moon is one sixth of what it is on Earth! This means that if you could find a place where the gravitational field strength was smaller than that of the Earth's you would lose weight without having to go through all that nasty and tedious dieting and exercise. You would still be the same shape sadly. It's just that you would weigh less!

The reality is that if you want to change your shape to something 'better' it is mass that you need to lose. Interestingly the indicator used by a number of bodies regarding obesity for example is the Body Mass Index (BMI) not the body weight index.

Your weight on other planets

The other planets in the solar system are of different masses and diameters from those of the Earth. As a result the gravitational field strength on the surface of these planets differs. Table 19.1 gives the values for the gravitational field strength (g) for some other planets.

Table 19.1

Planet	g
Mercury	3.6
Venus	8.8
Earth	10
Mars	3.8
Jupiter	26
Saturn	11
Uranus	11
Neptune	13

As you can see we could possibly live with the gravitational forces of most planets as they are only a little greater than 10 with Neptune making us about 30% heavier. Jupiter however would increase our weight to over 2 ½ times what it is on Earth. This has some implications for us:

- Our blood would be 2 ½ times heavier so our heart would struggle to pump it to our brain.
- Our bones would not be able to support our weight.
- Our head may be so heavy we would struggle to keep it upright.

On Mercury and Mars however we would weigh considerably less and that would cause other difficulties. Our normal walking style would be too strong and we would have to adopt a smaller shuffling style so as not to jump too high off the surface.

ACTIVITY

Calculating weight on other planets

Give the children a piece of card and ask them to draw a table with two columns similar to Table 19.1. In the second column calculate the weight of a 15kg animal on different planets. Ask learners to choose a planet.

They can then discuss the advantages and disadvantages of trying to survive in conditions of differing gravity. Asking them to discuss size, shape, number of legs etc. helps. They can then draw their animal in the space beneath the table as it would look on their chosen planet.

Consider a person in deep space far away from any large mass. That person would be virtually weightless. The scientists and astronauts in the International Space Station appear weightless, but this is due to the fact that the ISS is orbiting the Earth and they are in free fall. The gravitational attraction of the Earth still has a major effect at the distance of the ISS, but the orbit is what causes the 'weightlessness'.

Images of these astronauts and those in other shuttles etc. give the impression there is no gravity in space but this is untrue. It is the gravitational attraction from the Earth that holds them in orbit.

When we see images of astronauts undertaking an extra-vehicular activity to service a section of the space station for example, it looks as if they are just floating. The reality is that they are orbiting at something like 7.7 km/s (over 17,000 mph).

Photo 19.1

By NASA – https://en.wikipedia.org/wiki/Extravehicular_activity#/media/File:STS-116_spacewalk_1.jpg

Earth and the other planets orbit the Sun. Moons and satellites orbit the Earth. All of this is caused by gravitational attraction.

Other space objects such as asteroids or comets are generally moved through space because of some gravitational disturbance (a planet passing by for example). These objects will move through space until they are trapped by some object's gravitational field and pulled to the surface of that planet. Jupiter is considered to be a bit of a safety net for Earth. Its gravitational field is so strong it can attract many pieces of space debris and these crash into its atmosphere. The Shoemaker-Levy comet is a great example of this. In 1994 a large comet crashed into Jupiter. By the time it actually made contact it had broken into something like 21 pieces, each of which would have caused devastation on Earth had they reached us.

If we someday do travel to far-off planets the effect of living in little or no gravity is not trivial. Our bodies do not function as well in microgravity as they do on Earth. It may appear fun to float and push yourself off walls but this soon pales.

Not having any weight to push against weakens and decalcifies bones, particularly in the lower body. This release of calcium can lead to kidney stones and bone fractures. Muscles weaken as they no longer lift heavy limbs and astronauts spend time every day exercising to maintain muscle mass. Due to little or no gravity, fluid that would normally reside in the lower body is more evenly distributed and astronauts suffer from a slightly puffy face. The disks in astronauts' spines are not compressed and they get taller in space.

The lack of a daylight cycle and a confined ship can also lead to poor sleep and reduced work performance.

ESA and NASA spend a great deal of time and effort researching the factors that can alleviate these symptoms. If they are not successful in finding answers long space missions will never be possible.

On a more mundane yet still vitally important topic, normal everyday activities in zero gravity are very difficult.

Washing

Baths and showers do not operate as there is no gravity to pull the water down. The water forms little globules that move away when pushed or wiped. Astronauts use a damp cloth or sponge to just wipe themselves, and then use a towel to dry off.

Using the toilet

Urinating

Astronauts urinate into a hose via a personal funnel which is built for each astronaut and for their use alone! The hose has a vacuum at the other end and the fluid is gathered and passed to a recycling system which gets rid of various impurities and is then used for drinking water. The old Soviet Union space station, MIR, used to vent the urine into space, but the urine froze and cut through the solar panels and affected about 40% of its electrical capacity.

Defecating

This is a more problematic issue. A major problem is separation. With faecal matter being weightless it does not break off easily, and this caused great difficulty for early space travel such as for those astronauts who went to the Moon. The space station has a toilet with an airflow system and the astronauts train to sit on it perfectly, so that all matter is caught, stored in tanks, jettisoned to Earth, and then burns up on re-entry.

Gravity on television and in the cinema

It is very difficult to recreate micro gravity and weightlessness in a studio. *Apollo 13, 2001: A Space Odyssey* and *Gravity* tackled the problems relatively well, but most television series employ an 'artificial gravity' machine or something similar so that it appears as if there is normal gravity. This reduces the costs of having everything floating or moving about dramatically. It also gives the impression that prolonged space travel is possible when the difficulties are considerable. The images from television and cinema have a great impact on children but they often believe what they see. They can then retain these incorrect ideas or misconceptions and explain other examples using these.

Additionally, many movies involving space have sounds of things exploding or noise is made when a ship passes nearby. In many space movies ships are also destroyed in a ball of flame or explode in a similar fashion to things on Earth. The lack of any air/atmosphere means no sound will be transferred, and explosions generally involve objects being destroyed by the pressure of air rapidly expanding, and whilst this may happen in a spacecraft with a gaseous atmosphere this cannot happen in open space.

Films and television shows are great for stimulating debate and creating a sense of wonder, but not generally very good for reinforcing scientific principles.

Common misconceptions amongst children regarding gravity include the following:

* The Moon has no gravity.
* Planets with thin atmospheres have little gravity.
* Planets far from the Sun have little gravity.
* Astronauts in orbit are weightless.

The solar system

The solar system is generally defined as the area around the Sun where the Sun is the body with the greatest impact in terms of energy and gravity. This means the solar system extends far beyond Pluto. The Sun accounts for something like 99.9% of all the matter in the solar system, and Saturn and Jupiter account for more than 90% of the remaining mass.

The asteroid belt is an area of space between Mars and Jupiter composed of thousands and possibly millions of rocky objects. Even allowing for this the total mass of all the objects in the asteroid belt is less than one thousandth the mass of the Earth. Whilst books represent the asteroid belt as a dotty, filled with rocks region, the volume of space involved is so great that in reality it is very sparsely populated: the average distance between asteroids of say 1km across would be something of the order of 5–10 million km.

Gravity and tides

The gravitational attraction from the Moon and that from the Sun combine to cause some very regular effects on Earth. Whilst the Sun is an incredibly massive object the Moon is much closer. As a result the gravitational attraction of the Moon is roughly twice that of the Sun.

The Moon orbits the Earth and does so in a roughly circular orbit.

Figure 19.6

As the Moon orbits the Earth there is attraction between it and the section of Earth that is facing it. The sea is raised in this direction and the side facing the Moon has a high tide. The two sides of the Earth perpendicular to the Moon have a low tide, and the water on the far side of the Moon also has a high tide because of a complex explanation involving the Earth and Moon spinning.

In the course of one rotation of the Earth we generally have two high tides and two low tides. But the rotation of the Earth and the orbit of the Moon don't coincide exactly, so the times of the high and low tides are not the same every day.

There are also times when the Moon and the Sun are aligned with each other: at the time of the new or full moon they combine to create an extra high tide (and extra low tides at the side) called 'spring tides'.

There are also other times, roughly one week later, when the gravitational effects of the Sun and Moon act against each other to almost cancel each other out or to reduce the severity of the tides.

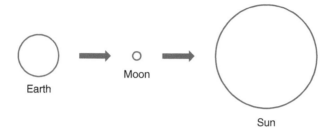

Figure 19.7

Figure 19.8

These moderate tides are known as 'neap tides'.

In reality there are number of other factors relating to the position and relative distances of the Sun and Moon which give rise to occasional very high and low tides, but it is an incredibly complex subject way beyond the scope of this text.

The purpose of this section on tides is not to give you as a teacher a deep understanding of the nature of tides, but to show that the gravitational attraction from these two bodies does combine to affect us on Earth.

Can the Moon's gravity affect children's behaviour?

Many rational people claim that the full moon triggers unusual behaviour patterns in people. The word 'lunacy' comes from diseases such as epilepsy which were thought to be caused by the Moon. It has no real bearing in science as the gravitational effects of the Moon on a person are so unbelievably small they can have no impact. A possible cause of bad behaviour relating to the Moon was thought to be the result of when people slept in open shelters where the full Moon kept them awake, and as a result they suffered from sleep deprivation. Sadly many teachers perpetuate this myth, but are more likely looking for some coincidence rather than any scientific analysis of bad behaviour.

The gravitational effects of other planets on the Earth are occasionally mentioned in newspaper stories of unusual astronomical events. The planets are so small in comparison to the Sun, and are so far away, that their effect is negligible. The alignment of the planets for example means nothing in terms of gravitational attraction to us – their effect is so small as to be virtually immeasurable.

Summary

This chapter outlined some of the key areas in the topic of weight, mass and gravity. It gave a brief explanation of the difference in terms' definitions as well as how this could be adapted to provide simple but appropriate explanations of the concepts as they may appear to learners. Whilst adults may appreciate the difference in the precise explanations of how things fall for example, children may find it difficult to overturn their original misconceptions. There is no simple way of addressing this. Many people will hold on to set beliefs in spite of evidence to the contrary.

REFLECTION POINTS

- How could you justify to a pupil that a dropped object will increase its falling speed (accelerate)?

    o This is best done experimentally. Take a small object such as a stone or a small ball and drop it from a range of heights. It may not be obvious that the ball is increasing in speed so you might need to use a bowl of sand or water to act as the landing zone. The size of the splash or the amount of sand displaced should be related to the motion of the ball and being dropped from a great height will splash more. This can be expanded to young children being able to jump off a bench but not being able to jump off a table or filing cabinet. Again this could be expanded to asking them to catch a heavyish object such as a bag of potatoes. They could catch this when it is released from a small height above them, but not if it is dropped from an upstairs window.

- Why does a stone fall more quickly than a piece of tissue paper when dropped from the same height?

    o When an object falls it does so because its weight pulls it down. As an object increases its speed there is greater air friction between it and the surrounding air. This friction stops it from increasing its speed. As the tissue is very light it does not need to move quickly for the air friction to have a noticeable effect. This is why it falls slowly. A stone has much greater weight and needs

(Continued)

(Continued)

to fall very quickly before the air friction has a noticeable effect. It will reach a terminal velocity but this will be much greater than that of a tissue. This leads to people believing heavy objects fall more quickly than light objects.

- List some of the possible benefits and disadvantages of prolonged space flight. (Given that it would take about two years to travel to Mars and back.)

  ○ Disadvantages: difficulties in washing, shaving and going to the toilet; loss of bone density and our heart becoming accustomed to not pumping very strongly; boredom; transport of two years' supply of food, water and oxygen; living with the same colleagues for so long; possible ill health (appendicitis?).
  ○ Advantages: many experiments may be undertaken to understand what life in weightless conditions is like, e.g. how do things burn? Can spiders make webs? No weight would reduce the chance of joint pain.

BiBLiOGRAPHY

Barker, S., Slingsby, D. and Tilling, S. (2002) 'Teaching Biology Outside the Classroom: Is it Heading for Extinction? A Report on Biology Fieldwork in the 14–19 Curriculum', Field Studies Council, British Ecological Society.

Chief Seattle, attributed to Chief Seattle (c. 1780–June 7, 1866).

Choudhary, N.L. (2012) *Chemical Properties & Analysis of Milk*. Uxbridge: Koros.

den Boer, L.C. and Schroten, A. (2007) 'Traffic noise reduction in Europe'. CE Delft.

Department for Education (2013) *National Curriculum for England: Framework for Key Stages 1 to 4*. London: DfE.

Dillon, J. et al. (2006) The value of outdoor learning: evidence from research in the UK and elsewhere, *School Science Review*, March, 87 (320).

Dobzhansky, T. (1973) Nothing in biology makes sense except in the light of evolution, *American Biology Teacher*, 35(3): 125–9. Available at www.tiny.cc/dobzhansky-1973

Education Scotland (2011) *Outdoor Learning: Section 5: Places to Learn Outdoors: Using the Local Area*. Available at www.educationscotland.gov.uk (last accessed 3 September 2016).

Goldsworthy, A., Feasey, R. and Ball, S. (1997) *Making Sense of Primary Science Investigations*. Hatfield: Association for Science Education.

Harlen, W. (ed.) (2010) *Principles and Big Ideas of Science Education*. Hatfield: Association for Science Education. Available at www.interacademies.net/File.aspx?id=25103

Haslam, F. and Gunstone, R. (1996) 'Observation in Science Classes: Students' Beliefs about Its Nature and Purpose'. Paper presented at the 69th Annual Meeting of the National Association for Research in Science Teaching (St. Louis, MO, April).

Learning and Teaching Scotland (2010) *The Curriculum for Excellence through Outdoor Learning*. Glasgow: Learning and Teaching Scotland.

Nundy, S. (2001) *Raising Achievement through the Environment: A Case for Fieldwork and Field Centres*. Peterborough: National Association of Field Studies Officers.

Oliver, A. (2006) *Creative Teaching: Science in the Early Years and Primary Classroom*. London: David Fulton. Available at http://web.primaryevolution.com/?page_id=350 (last accessed 8 October 2015).

Outdoor Science Working Group (2011) 'Outdoor Science: A Co-ordinated Approach to High-quality Teaching and Learning in Fieldwork for Science Education', Association for Science Education/Nuffield Foundation, Field Studies Council Occasional Publication 144.

Research Defence Society (2005) 'Declaration on Animals in Medical Research', London, Research Defence Society.

Sefton, I. (2004) 'Understanding energy'. Available at science.universe.edu.au/school/curric/stage6/phys/stw2004/sefton1.pdf (last accessed 02 February 2017).

Souter, N. and Lewthwaite, K. (2007) *Nature in a Changing Climate: Phenology Uncovered.* Grantham: Woodland Trust and Association for Science Education.

Woodland Trust (2017) Welcome to Nature's Calendar (online), WT. Available at http://www.naturescalendar.org.uk/survey/ (last accessed 13 January 2017).

INDEX

Photos are indicated by page numbers in bold print.

adaptation 2, **3**, 112, 116–17, 156

air:
  and cloud formation 201
  composition 196
  compressed 197–8
  friction and drag 350–3
  as natural resource 229

aircraft noise 334

alcohol 76

algae 12, 13, 37–8

alloys 238–9

aluminium 199

amoeba 122

anaerobic digestion 85

animal testing 166

animals:
  behaviour:
    chemical communication 174–5
    habituation 175–6
    innate 172
    learned 172
    response to the environment 178–9
    social 172–3
    sound 173–4
    types of 171
    visual signals 174
  breathing 125–7
  circulatory system 138–41
  classification 17–18
  in the classroom 170–1
  damaged tissue and repair 73
  digestion 127–32
  evolution 3–4
  extinction 24–5
  identification keys 155
  introductions into habitats 50, 51
  invertebrates ('minibeasts') 18–20, 43–5, 136–7
  life cycles 162–3
  migration 51, 173

animals *cont.*
  as natural resources 229
  reproduction 142, 156
  senses 180–6
  signs in soil **37**, **38**
  skeleton 135–8
  using electricity 124
  variation 158–9
  see also individual species

anthracite 230

ants 174–5

Apollo 13 (film) 368

argon 196

Aristotle 342

artemia (brine shrimps) 163

arteries 139

aspirin 95

astronauts **366**–8

astronomy *see* early astronomy; space

atoms 237, 294

bacterial industries 78–80
  biological washing powders 79
  sauerkraut and kimachi 79
  silage 78

Banks, Joseph, 150

barnacle geese 173

bees 172–3, 174

bicarbonate of soda 243, 245–6

Big Garden Birdwatch 179

biodiesel 85

biodiversity 2–26
  collecting and sorting 5–11, 18–24
  evolution and adaptation 2–4
  exdinction 24–5
  and genetic diversity 3, 147
  identification 11–18, 25
  out-of-classroom learning 4–5, 7–8
  see also animals

biofuels 95

biogas 84, 85, 255
biomass 34, 84
biorenewables 84–5
birds:
   habituation to scarecrows 175
   migration 173
   observation of 178–9
   pellets **38**
bison 51
bituminous coal 230
blood 135, 138–9
body:
   brain 180
   breathing 125–7
   cells 156
   circulatory system 138–41
   digestion 127–32
   drinking 250–1
   elements from stellar reaction 267
   senses 180–6
   skeleton 135–8
   sweating 257
   waste 83–4, 255
   weightless in space 367–8
   see also animals; cell biology
body awareness 123
boiling 200–1
bones *see* skeletons
brain 180
bread 76–7, 81–2
breathing 125–7, 240
brewing 76
burning 237, 238–40, 284
   food 240
butterflies **129**

calendar 273, 275, 280
calories 282, 283
car brakes 195
carbon cycle 64
carbon dioxide:
   as component of air 196
   as compound 237
   in fire extinguishers 243
   as 'greenhouse gas' 231
   in photosynthesis 110, 111, 113, 117
carbon monoxide 239
carnivores 47
'carousel brainstorming' 141
caterpillars 129
cell biology 58–86, 147
   bacterial industries 78–80
   blood cells 135, 139–40
   cell chemistry 62–4, 76

cell biology *cont.*
   cell theory 59
   division 72–3
      cancer 75
      growth 74–5, 96, 105
      tissue culture **73**
      tumours 74–5
   environmental biotechnology 80–6
   enzymes 80
   eukaryotes 60
   food 64–72
   gene therapy 80
   prokaryotes 60
   sex cells 61
   stem cells 75, 75–6
   using digital microscopes **62**
   yeast 76–8
chemical changes 67, 236–46
   alloys 238–9
   burning 237, 238–40
   experiments 243–6
   in food 241–3
   reactions with water 241
   respiration 240
chemical symbols 237
Chinchona bark 95
chlorine 253
chromosomes 156
circulatory system 138–41
   heart 140–1
citizen science 162, 179, 180
climate change 51, 166, 231–2
clinometers 35
clouds 201, 257–8
   seeding 257
coal 230
Coco de Mer **98**
collecting 148–55
   and classification 150, 151–5
   and identification 150
   plant materials 150–2
Colorado beetles **50**–1
combustion 239
   see also burning
communication of animals 173–4
compasses 349
compounds 237
Concorde 192
conifers 12, **16**, 91
conservation 12, 24–5, 30–1
   and out-of-classroom learning 47–8
Cook, Captain, 150
Copernicus 343
copper 199, 238, 241, 302

country parks 30
Crick, Francis 147, 167

daffodils 105
Darwin, Charles 2, 147, 150
Dead Sea **212**, 249, **250**
decomposers 47, 80
decomposition 80–3
dehydration:
    of body 184, 250, 251, 261
    for food preservation 68
density 211–13
deserts 117
digestion 127–32
    chemical breakdown of food 132
    digestive system 130
    physical breakdown of food 128–31
        chopping food 128–9
        mixing food 130–1
        moving food 130
    polysaccharides 131
digital microscopes 62
dinosaurs 24
diseases 49, 80, 147, 256
DNA 10, 64, 75, 80, 146–8
Dobzhansky, T. 3

early astronomy 270
    and astrology 270–1
    planetary theories 271–2
ears and hearing 181, 183–4, 327–31
    headphones 330, 339
    hearing aids 337–8
    hearing loss 330
    loudness 328–9, 330–1
    structure of ear 327–8
    tinnitus 330
Earth:
    climate change 231–2
    formation 222
    gravity 358, 365
    magnetism 349
    in space 268, 269, 271–8
        and Moon 275, 369
        orbit 272–5, 275–6, 279
        rotation and length of days 275–7, 279
        seasons 274–5
        and tides 369–70
        year 273–4
    structure 222–4
        crust 222, 223
        inner core 223
        mantle 223
        outer core 223

Earth *cont.*
    surface structure 224
    temperature of core 223–4
    see also fossil fuels; fossils; rocks; soil
earthquakes 224
earthworms 125, 130, 177–8
    movement 137
ecosystems 28–56
electricity 294–307
    in animals 124
    batteries 289–90, 306
    cables and pylons 302–**3**
    charged objects 295–9
        balloon experiment 295
        discharging 296
        everyday objects 296–7
        negative and positive charge 295
        shocks 297–9
        triboelectric 295
    circuits 289–90, 299–301, **300**, 307
        parallel circuits 304–5
        series circuits 303–4
        supply voltage 301
        symbols 300–1
    current 301–2
    generating electricity 305
    insulators 302
    National Grid 289, 305
    and 'power' 290
    safety 305–6
    watts 290
    wiring 306
electrons 294
elements 237, 267
    in ancient Greek philosophy 249
energy 282–92
    batteries and electrical energy 289–90
    children's views 283–4
    conservation 124, 285, 290–1
    energy efficiency 291
    everyday uses 285
    from food 240, 282, 283
    forms 284–5
    heat 286–9
    measurement 282–3
    in nature 124–5
    in photosynthesis 112
    from the Sun 267, 285
    transfer in water cycle 256–7
    wasted 291
environmental education:
    resources 31–2
    samples and estimates 33–6, 41–5
    trails 32–3

enzymes 132
equinoxes 275
Eratosthenes 271
evaporation 202–3, 217–19, 257–8
evolution 2–4, 123, 147, 150
   controversy 3–4, 164
   descent with modification 3
   plants 110
extinction 24–5
eyes and vision 181–3, 310–13, 315–17
   colour vision 316–17
   eye structure 311–12
   field of view 315
   lenses 321, 324
   observing the Sun 314
   pinhole camera 312–13
   visual acuity 316

fermentation 95
ferns **15**, 90–1
Feyman, Richard 165, 284
fieldwork sites 29–30
fluoride 254
food:
   additives 134–5
   canned food 68
   cereals 94, 132–3
   cheese 69, 79, 80
   chemical changes 241–3
   colour change 241–2
   cooking:
      and chemistry 67
      edibility and palatability 69
      and heat transfer **288**–9
      popcorn 66–7
      purpose of cooking 66–7
   decomposition **80–1**
   dehydration 68
   diet 133–4
   energy from 240
   fermentation 71
   flavour 69
   flowers 94
   food chains 45–7
   food webs 47, 50, 53–4, 110
   fruits 94
   labelling 64–5
   leaves 94
   milk 69–72
      cheese 71–2
      curds and whey 70
      lactose mutation 69–70
      pasteurisation 71
      sour milk 243
      yoghurt 71
   mould 69

food *cont.*
   nutrition 133–4
   oils 94
   pickling 66
   preservation 68, 69
   pulses 133
   refrigeration and freezing 68
   roots 94
   sap of plants 94
   seeds 94
   and sense of taste 185–6
   silage 78
   staple foods 132–3
   waste 84
   yeast 69, 76
      bread 76–7
      ginger beer 77–8
forces 342–56
   changes:
      direction 346
      shape 345–6, 349–50
      speed 343–4, 349
   at a distance:
      gravity 346, 358–72
      magnets 347–9
   friction 349–53, 361–3
   identifying 353–4
   machines 354–6
   measuring 344–5
   and motion: ancient Greek ideas 342–3
fossil fuels 229–33
   effects of utilisation 231–2
   energy conservation 233
   formation 229–30
   reduction of use 232, 233, 290
fossils 227–9
   making a fossil 227–8
Franklin, Rosalind 147
freezing 200
freshwater 117, 218–19
friction 349–53
   air friction and drag 350–3
      aeroplanes 350
      and air resistance **352**–3
      cars 350–**1**
      falling objects 361–3
      streamlining **353**
Friends of the Earth 25
frog life cycle 125
fungi **36**, **52**, 69
   antibiotics 79
   see also mould

galaxies **265–6**
Galileo 343, 359

gas (fossil fuel) 230
gas lights 315
gases 196–8
  compression 197
  expansion and contraction 198
genetics
  DNA 10, 64, 75, 80, 146–8, 156
  gene therapy 80
  genetic modification 74, 147, 164
  genomes 147
  inheritance 155–8
  inherited diseases 147
  predictive genetics 156–8
  profiles 147
  variation 3, 158–62
    continuous 159
    discrete/discontinuous 160
glass 321
gold 217, 238, 302
grass germination 104
gravity 346, 358–72
  common misconceptions 368
  falling objects 359–63, 371–2
    air friction (drag) 362–3
    heavy and light objects 360–1
    speed of fall 359–61, 361–2
    terminal velocity 362–3
  solar system 368–9
  on television and film 368
  tides 368–70
  weight and mass 363–8
    and astronauts in space 366–8
Gravity (film) 368
Greenpeace 25
gymnosperms 91

habitats 48–53
  abiotic factors 48
  and adaptation 116–17
  biotic factors 48–9
    competition 49
    disease 49
    food 49
    predation 49
  changes 50–2
    addition and removal 52–4
    populations 51–2
hardness 209
hawthorn 11–12
health:
  of an environment 51
  and diet 71, 133–4
  and drinking water 256, 261
  of plants 64
  in space flight 367–89

health and safety
  electricity 305–6
  out-of-classroom learning 29, 47–8
heart 140–1
heat 286–9
  burning 237, 238–40
  changes of state 67, 192, 195–6, 198–203, 259
  conservation 286
  convection 288
  in cooking **288**–9
  and dissolving 215
  in the Earth's core 223–4
  heat 'rising' 287
  'hot' and 'cold' 286, 287
  sources 288–9
  tranference 286
helium 267
Heraclides 271
herbivores 46
Himalayan balsam (*Impatiens glandulifera*) 51
Himalayas 224
Hooke, Robert 86, 345, 358
hornworts 89
horsetails **14**–15, 91
Human Genome Proect 147
human life cycle 142–3
hydraulics 195
hydrogen 199, 267
hypotheses 165

icebergs 259–60
identification and classification 9–24
  animals 17–18
  databases 11
  'I-SPY' approach 41
  living and non-living 10–11
  'minibeasts' 43
  'new' species 25
  plants 11–17, 23
  resources 40–1
inheritance 3, 155–8
  genetic information 155–6
insects 33–4, 163
  see also invertebrates ('minibeasts')
interdependency of living things 28
invertebrates ('minibeasts')
  body types 137
  collecting and keeping 18–20, 43–5, 171
    hand lens (loupe) **21**
    pitfall traps 20, 45
    pooters 43, 45
    sweep nets 20–1, 44–5, **44**
    typical species 43
  earthworms 177–8
  exoskeletons 136–7

invertebrates ('minibeasts') *cont.*
    identification keys 155
    life cycles 163
    in soil 37
    surveys/sampling 43–5
        ground dwelling 44–5
        trees and shrubs 33–**4**, 43
'invisible ink' 245–6
iron 238, 241, 347, 348

Jenkins, A.J.: 'Animal Sounds' 184
Joule, James Prescott 282
Jupiter 268, 270, 365, 367, 369

Kepler, Johannes 359
keys for identification 152–5
    making 154–5
kidneys 250
kinetic energy 199
Kipling, Rudyard: 'I Keep Six Honest Serving-Men' 149

larch cone **16**
lava 225
lava experiment 244–5
lead 238
Learning and Teaching Scotland 29
lenses 321
life cycles:
    of frog 125
    human 142–3
    of living things 162–3
    of plants 96–7, 100
life processes 122–44
    circulatory system 138–41
    digestion 127–32
    human life cycle 142–3
    respiration 63, 123–7, 240
    support and movement 135–8
    see also food
light 310–24
    eyes and vision 181–3, 310–13, 315–17
    lenses 321, 324
    in photosynthesis 110–11, 112, 310
    pinhole camera 312–13
    from Sun 313–14
    reflected 314, 318–**20**
    shadows 322–3, 324
    street and domestic lighting 315
    transmission in materials 320–1
        transparent materials 321
    travelling 317–18, 324
    Ulta Violet light 313
lightning 297

lignite 230
liquids 193–6
    evaporation 202–3
    hydraulics 195
    and temperature 195–6, 200–1
    vapourisation 200–1
    see also water
liver 131
liverworts 89
living and non-living things 10–11, 122, 149

magma 225
magnets 347–9
    industrial magnets 348
    uses 349
marble **226**
Mars 268, 270, 365
materials:
    conductors and insulators of electricity 302–3
    from the Earth:
        fossil fuels 229–33
        rocks 224–6
    magnetic materials 347–8
    from plants 93–6, 229
    physical properties 208–19
        colour 208
        density 211–13
        hardness 209
        intensive and extensive 208
        mass 208
        solubility 213–19
        strength 209–**11**
    transparent and translucent 320–1
    see also states of matter
medicine 94–5
melting 198–200, 237
    melting points 199
mercury (metal) 195, 199
Mercury (planet) 268, 270, 365
metabolism 123
    of plants 112
metals 199, 238
    attraction to magnets 348
    as conductors of electricity 301–2
    for jewellery 237–8
    and rust 241
metamorphosis 163
migration 51
milk 69–72, 243
Milky Way **266**
Mimosa pudica **116**
minerals 229
mirrors 318–20

mixtures:
  homogenous and heterogenous 213–14
  separating 215–17
    filtering 216
    panning for gold **217**
    settling 216
    solid and liquid mixtures 216, 217
    solid mixtures 216
moecules 237
molecular biology 147
Moon 222, 267, 269, 279, 314
  affects on behaviour 370–1
  in early astronomy 270
  and month 275, 280
  and tides 369–70
  weight of objects on 364
mosses 13–14, 89, 90, 152
mould 79, 201
mountains 224
Murdoch, William 315
muscles 138, 139
  for eye movement 181
music 332–4
  percussive instruments 332
  stringed instruments 333–4
  wind instruments 332–3

national parks 30
Natura Sites 31
natural resources 228–9
Nature's Calendar 179
Neptune 268, 365
nettles **117**–18
Newton balances **344**, 345
Newton, Sir Isaac 314, 343, 345, 358
Newtons (N) 363, 364
nickel 199
nitrogen 196
nose 181
nuclear fusion 267
nuclear power 166
nucleus 294

observation skills 6–9
  137-second plant hunt 8–9
  and collecting 148–55
  familiar objects 6–7
  and identification 152–3
  listening to the heart 140–1
  looking closely ('binoculars') **8**
oil (fossil fuel) 230
oils from plants 94, 95
omnivores 47

Orion (constellation) **264**
out-of-classroom learning:
  advantages 4–5, 28–9
  conduct and conservation 47–8
  health and safety 47–8
  local resources 5, 8
  nature reserves 30
  parks 30
  planning 47–8
  school grounds 5, 29
  trails 32–3
Outdoor Science Working Group 29
owls 176–7
oxygen:
  and burning 239–40
  as component of air 196
  in photosynthesis 110

pancreas 131
paper 210
parachutes 344
Pasteur, Louis 71
peat 90, 230–1
perfumes 203
periodic table 237, 238
periscopes 319
phenology 51
pheromones 174
photosynthesis 63–4, 110–11, 112, 229, 310
pinhole camera 312–13
planets 264, 267–73, 314
  names 270
  weight of objects on 365–7
  see also individual planets by name;
    solar system
plantains **41**
plants:
  137-second plant hunt 8–9
  adaptation **3**, 88, 112, 116–17
  biorenewables 84–5
  broad bean life cycle, **97**
  cell division and tissue culture **73**, 105
  cereals 94
  chlorophyll 111–12
  classification 88
  cloning 74, 107
  for clothing 94
  collecting materials 22–4, 150–2
    leaf rubbings 23
    leaf skeletons 23–4
    pressed flowers 24, 152
    pressed leaves 23
    winter twigs 150–1

plants *cont.*
  compost 82–3
  defences 117–18
  definition 88
  desert 117
  flowering (*angiosperms*) 16–17, 92
    dicots and monocots 92
    flower dissection 109
    sexual reproduction 108–9
  as food 94
  freshwater 117
  fruits 94, 99, **100**
  gas production 110
  genetic modification 74, 86
  germination 72, 97, 102–3, 104, 108
  growth 73–5, 96, 97–8, 114
  identification 11–17, 23, 151
    keys 152–5
  introductions into habitats 51
  leaf veins 93
  leaves 94, 111, 112, 151
    classification 153–4
  life cycle 96–7
  lifespan 105
  materials/resources from 93–6, 229
  medicine 94–5
  meristem 105
  millennium seed bank 12
  non-flowering 12–16, 89–91
  oils 94, 95
  photosynthesis 63–4, 89, 93, 229, 310
    chemistry of 110–12
  pollination **28**
  safety 118
  and sea (salt) water 250
  seeds 72, 94, 96–105
    dispersal 99–102
    germination 102–5, **103**
    structure **98**–9
  sensitivity 114, 180
  stems 112
  surveying 41–3
  variation 89, 158–9, **162**
  vegetative propagation 105–8
    artificial processes 106, 107–8
    bulbs 105
    carrot tops **107**
    and differentiation 106
    micropropagation 106
    runners 106
    tubers 105
  water transport 102, 112–**13**
  see also food; *individual species*

plasma 190
Pluto 267
Polytrichum moss **115**
pond weeds 110
poppies 49, **50**
potassium 199
potatoes 105
power 290
producers 46, 110
protected ecosystems 30–1
proteins 69, 131
Ptolemy 271
pumice stone **225**

quadrats **42**
quorn 69

rabbits 51
rainfall 257
  in Britain 252
red deer 49, 173
regional parks 30
reservoirs **252**
respiration 63, 123–7, 240
  aerobic 124
    and anaerobic 63
  breathing 125–7, 240
  energy transformations 124–5
  gas exchange 125
restriction enzymes 147
robins 172
rocks 224–6
  as building materials 229
  igneous rocks 224–5
  metamorphic rocks 225–6
  sedimentary rocks 225
  see also fossils; soil
Rose, Francis: *The Wildflower Key* 152
Royal Botanical Gardens 12
rubber 95, 96
rust 241

salt:
  for food preservation 68
  salt pans **218**
  sea (salt) water 211–13, 218, 249–51
  solubility 213
samples and estimates 33–6, 41–5
  clinometers 35
  estimating height 34–6
  invertebrates 43–5
  plant populations 42–3
  quadrats **42**

sandstone **226**
satellites 269–70, 314
Saturn 268, 270, 365
school wildlife areas 5
science:
    and controversy 3–4, 165–6
    and media reports 166–7
    and religion 4, 163–4
    scientific knowledge 165
SCUBA divers 197
sea (salt) water 218, 249–51
    density 211–**12**
seasons 274–5
senses and sense organs 180–6
sewage 83
sex cells (gametes) 61
sex chromosomes 156
sexual reproduction 61
sharks **352**–3
silver 238
Sites of Special Scientific Interest (SSSI) 31
skeletons 135–8
    exoskeletons 136–7
    and muscles 138, 139
    no skeletons 137
skin 181, 184–5
    and sensitivity 184–5
skydivers 262, 263
snails 125, 129, 175–6
snow 258–9
Sodeberg, Patti 158
soil 36–41, 226–7
    abiotic factors 40
    biotic factors 36–9
    examining textures 40
    formation 226–7
    identification of specimens 40–1
    invertebrates ('minibeasts') 37
    signs of animals **37**
    see also rocks
solar system 264, 267–9, **268**
    asteroids 367, 369
    Earth's orbit and rotation 272–9
    geocentric model 271–2
    gravity 368–9
    heliocentric model 272–3
    names of planets 270–1
    scale representation 268–9
solids 191–3
    contraction 192–**3**
    definition 191
    expansion **192**, 193, 259
solstices 275

solutions 214–15, 237
    separating 217–19
    and temperature 215
    see also mixtures
sound 326–39
    cause 326
    chimes 331
    decibels 328–9
    ears and hearing 181, 183–4, 327–31
        hearing aids 337–8
    frequency (pitch) 327, 328
    making sounds 332–4
    music 332–4
    and noise 326–7, 331–2, 338
    noise pollution 334–5
    sound propagation 335–7, **336**
        amplification 336, 337
    vibrations 338–9
space:
    astronauts **366**–8
    early astronomy 270–3
    stars and galaxies 264–6
    Sun and solar system 266–70
    universe: definition 264
    years, months and days 273–8
    see also Earth
Sphagnum moss 90
spider plants 106
sporophytes 14
stars:
    in astrology 271
    constellations **264**–5
    as plasma 190
states of matter 190–203
    changes:
        boiling 200–1
        condensing 201
        evaporation 202–3
        freezing 200, 259
        melting 198–200
    gases 196–8
    liquids 193–6
    solids 191–3
steel 238
stem cell research 107
stick insects 163
strength of materials 209–11
    shape 210–**11**
strontium 238
Sun 264, 266–**7**, 369
    charting position of 277–8
        over course of the year **278**
    and length of days 275–7
    observing the Sun 314

Sun *cont.*
    radiating light 313–14
    and seasons 274–5
    sunlight as natural resource 229
    in water cycle 257
    see also Earth; solar system; stars
sundials 322–3
sustainable resources 229
sycamore 101

taste and smell 185–6
taxonomy 152
teeth 128
    dental care 129
temperature 198–9
    global warming 231, 232
    and length of days 276–7
    regulated by the body 203
    and solubility 215
theories in science 58, 59, 165
thermometers 195–**6**
tin 199
Tomlinson, Jill: *The Owl That Was Afraid of the Dark* 176
tongue 181
traffic noise 335
trees 33–4
    conifers 91
    estimating height 34–6
    surveys 42
    timber 95, 96
tungsten 199
turtles 172
2001: A Space Odyssey (film) 368

UN Framework Convention on Climate Change (UNFCCC) 232
UN General Assembly statement on water 248
UN Millennium Development Goals 30–1
universe *see* space
Uranus 267, 268, 365

veins 139
Venus 268, 270, 365
Venus fly trap 115–16
viruses 80
vision *see* eyes and vision

Wallace, Alfred Russel 2, 3
waste:
    decomposition 80–3
    water 83–4

water 248–61
    boiling 200–1
    chemical reactions with 241
    condensation 201
    conservation 261
    density 211
    drinking 250–1
    drinking water 251–6
        across the world 256
        collection **252**
        disinfection 253–4
        particle removal 253
        screening **253**
        treatment system 254–5
    evaporation 202–3, 217–19, 249
    filtering 84, 253
    floating 211–13, 249
    freshwater 117, 218–19, 248–9
    ice 200, 258–60, **259**
    as liquid 193–4
    as natural resource 229
    sea (salt) water 211–13, 218, 249–51
        for irrigation 250–1
    steam 202
    waste 83–4, 255
    water cycle 256–8
    waterborne diseases 256
water vapour 196, 201, 257
Watson, James 147, 167
Watt, James 290
weather:
    climate change 51, 166, 231–2
    clouds 201, 257–8
    effects of global warming 232
    frost 260
    lightning 297
    rainfall 257, 258
    snow 258–9
weight and mass 363–8
    mass 364
    on other planets 365–7
    and 'weightlessness' in space 366–8
Wilkins, Maurice 147, 167
willow 95
witch's broom gall **74**
wood pulp 95
World Health Organisation (WHO) 256
World Wildlife Fund for Nature (WWF) 24, 25

Yellowstone National Park 30